Springer INdAM Series

Volume 54

This series will publish textbooks, multi-authors books, thesis and monographs in English language resulting from workshops, conferences, courses, schools, seminars, doctoral thesis, and research activities carried out at INDAM - Istituto Nazionale di Alta Matematica, http://www.altamatematica.it/en. The books in the series will discuss recent results and analyze new trends in mathematics and its applications.

THE SERIES IS INDEXED IN SCOPUS

Emiliano Cristiani • Maurizio Falcone •
Silvia Tozza
Editors

Mathematical Methods for Objects Reconstruction

From 3D Vision to 3D Printing

Springer

Editors
Emiliano Cristiani
Istituto per le Applicazioni del Calcolo
"M.Picone"
Consiglio Nazionale delle Ricerche
Rome, Italy

Maurizio Falcone *(deceased)*
Department of Mathematics
Sapienza University of Rome
Rome, Italy

Silvia Tozza
Department of Mathematics
University of Bologna
Bologna, Italy

ISSN 2281-518X ISSN 2281-5198 (electronic)
Springer INdAM Series
ISBN 978-981-99-0775-5 ISBN 978-981-99-0776-2 (eBook)
https://doi.org/10.1007/978-981-99-0776-2

This Springer imprint is published by the registered company Springer Nature Singapore Pte Ltd.
The registered company address is: 152 Beach Road, #21-01/04 Gateway East, Singapore 189721, Singapore

To Maurizio,
our Academic Father

Preface

Three-dimensional (3D) reconstruction of the shape of objects is an issue that has been investigated largely by the computer vision and applied mathematics communities since the last century. The class of problems related to that issue is the so-called Shape-from-X class, where the X specifies the kind of data used for the reconstruction (e.g., shading, texture, template, polarization).

The main Shape-from-X techniques can be classified as photometric or geometric based, and/or in relation to the number of images they require. Geometric Shape-from-X techniques are built upon the identification of features in the image(s). An example is the Shape-from-Template (SfT), which uses a shape template to infer the shape of a deformable object observed in an image. On the other hand, photometric Shape-from-X techniques are based on the analysis of the quantity of light received in each photosite of the camera's sensor. An example can be the classical Shape-from-Shading problem, formalized in that sense since the pioneering work by B.K.P. Horn in the 1970s, which utilizes as a datum a single 2D gray-level image of the object to be reconstructed.

All these 3D reconstruction problems are typically ill posed since they do not admit, in general, a unique solution. Hence, advanced techniques for the analysis and for the numerical approximation and/or a priori knowledge are required.

Most of the problems belonging to the Shape-from-X class can be formulated via partial differential equations and/or via variational methods, giving rise to a variety of nonlinear systems that have been analyzed by many authors.

Shape-from-X problems have a natural counterpart in 3D printing, which consists in producing a 3D printed object with desired appearance and physical properties. Nevertheless, the two areas have received different attention from the computer vision and applied mathematics communities, Shape-from-X being much more investigated than the other. The mathematical aspects of 3D printing have begun to be explored only since 2017. So far, research has focused mainly on the shape optimization of overhangs, those temporary structures that must be printed to support the actual object during construction and then removed at the end of the process. To this end, long-standing and mature mathematical tools for shape optimization have been adapted to this new application, leading to great results (especially for saving printing material). In particular, we are referring to those mathematical tools based on elastic displacement, Hamilton-Jacobi equations, front propagation problems, and level set method.

Only recently, mathematical research has expanded somewhat in order to include optimal path generation of the extruder (important for reducing printing time), minimization of thermal stress (important for the quality of the final product), and optimal object partitioning.

One of the most interesting problems that fully relates 3D vision to 3D printing is probably the appearance replication. This problem, only partially explored, consists in replicating (multi-material) real objects with complex reflectance features using a single, cheaper printing material, possibly with the simple diffuse Lambertian reflectance. To trick the eye, the surface of the object is rippled with tiny facets that regulate the reflection of light, analogous to what is done, for example, in the Oren-Nayar reflectance model for recovering the 3D shape of the object in the context of the Shape-from-Shading problem.

This volume is devoted to mathematical and numerical methods for object reconstruction, and it aims at creating a bridge between 3D vision and 3D printing, moving from the 3D data acquisition and 3D reconstruction to the 3D printing of the reconstructed object, with software development and/or new mathematical methods to get closer and closer to real-world applications. Some contributions focus on 3D vision, dedicated to photometric- or geometric-based Shape-from-X problems. Other contributions address specific issues related to 3D printing, further widening the research topics of this newly investigated area.

This book is useful for both academic researchers and experts from industry working in these areas who want to focus on complementary knowledge in 3D vision and 3D printing fields. Also practitioners and graduate students working on partial differential equations, optimization methods, and related numerical analysis will find this volume interesting as an approach to the field.

The research contributions contained in the book give only a partial overview of the research directions and various techniques of heterogeneous origin discussed during the INdAM Workshop "Mathematical Methods for Objects Reconstruction: From 3D Vision to 3D Printing", held online due to COVID-19 pandemic, February 10–12, 2021.

We want to thank all the speakers at that workshop and those who contributed to the present volume, which we hope will attract new researchers to this challenging area.

A particular mention goes to Maurizio Falcone, our Academic Father, who disappeared suddenly and prematurely in November 2022, during the publication process of this book. Maurizio followed carefully almost all the steps of this book, as co-editor and also as co-author of the first Chapter. We thank him for what he did for us and in general for his contributions to Applied Mathematics. We miss him, a feeling shared by all the authors of this book, colleagues and friends. He leaves an unfillable void in our lives.

The following is a brief description of the chapters contained here.

In Chap. 1, Cristiani et al. present a brief overview of techniques for 3D reconstruction (solving the classical Shape-from-Shading, Photometric Stereo, and Multi-view Shape-from-Shading problems), and some issues for 3D printing, for example, overhang and infill, emphasizing the approaches based on nonlinear partial differential equations and their numerical resolution. The goal is to present introductory materials to readers, stressing the common mathematical techniques and the possible interactions among the two areas, which nowadays are still limited. The chapter also presents two appendices on STL and G-code file formats, which are largely used in 3D printing and are useful to create a final 3D print from scratch without relying on existing compiled software.

In Chap. 2, Rodriguez et al. focus on uncalibrated photometric-stereo problem using non-Lambertian reflectance models. In more detail, Hayakawa's procedure is used to detect light positions, provided that at least six images of the sample object with different lighting directions are available, and the Oren-Nayar model is used as preprocessing for tuning Hayakawa's detected light directions. The authors stress the importance of the roughness parameter in estimating the light directions and in reconstructing the object. They determine a realistic range of variation for the roughness parameter, which results in a set of meaningful 3D reconstructions in an outdoor environment.

In Chap. 3, Collins and Bartoli tackle the SfT problem, aiming to reconstruct the 3D shape of a deformable object from a monocular image. The authors present a novel SfT method that handles unknown focal length, calling it fSfT. The fSfT problem is solved by gradient-based optimization of a large-scale nonconvex cost function, which requires a suitable initialization, and cost-normalization strategies are presented, allowing the same cost-function weights to be used in a diverse range of cases. Numerical results on 12 public datasets are reported in order to show the effectiveness of the proposed fSfT method in both focal length and deformation accuracy for real-world applications.

In Chap. 4, Bærentzen et al. focus on 3D reconstruction from a discrete set of data points in space, that is, a point cloud. Specifically, they are interested in 3D reconstruction of real-world objects, with very thin tubular structures, which are hard to reconstruct using traditional methods. The proposed procedure constructs a skeleton of the object from a graph, whose vertices are the input points. Then, a surface representation is created from the skeleton, and, finally, a triangular mesh is generated from the surface representation. Following this pipeline, they are able to reconstruct a valid surface model from the data. The results demonstrate the efficacy of the proposed method on a tree acquired using ground-based LiDAR.

In Chap. 5, Beisegel et al. identify a novel optimization problem in both wire arc additive manufacturing and powder bed fusion 3D printing processes. The problem belongs to the class of mixed integer programming techniques with partial differential equations as a further constraint. An important novelty here is that heat transfer is taken into account in the optimization process, aiming at lowering the internal thermal stress of the object.

In Chap. 6, Yarahmadi et al. deal with the powder bed fusion printing process and propose an optimization heuristic to find the optimal laser beam trajectory. The devised optimization procedure, as in the previous chapter, takes into consideration the thermal stress of the object in order to minimize the average thermal gradient.

Rome, Italy
Bologna, Italy
December 9, 2022

Emiliano Cristiani
Silvia Tozza

Contents

An Overview of Some Mathematical Techniques and Problems Linking 3D Vision to 3D Printing

Emiliano Cristiani, Maurizio Falcone, and Silvia Tozza

Abstract Computer Vision and 3D printing have rapidly evolved in the last 10 years but interactions among them have been very limited so far, despite the fact that they share several mathematical techniques. We try to fill the gap presenting an overview of some techniques for Shape-from-Shading problems as well as for 3D printing with an emphasis on the approaches based on nonlinear partial differential equations and optimization. We also sketch possible couplings to complete the process of object manufacturing starting from one or more images of the object and ending with its final 3D print. We will give some practical examples of this procedure.

Keywords Shape-from-shading · Photometric stereo technique · Multi-view SfS · 3D vision · 3D printing · Overhangs · Infill

1 Introduction

In general, the Shape-from-Shading (SfS) problem is described by the so-called *irradiance equation*

$$I(x, y) = R(\mathbf{N}(x, y), \mathbf{\Lambda}). \tag{1}$$

Maurizio Falcone died before publication of this work was completed

E. Cristiani
Istituto per le Applicazioni del Calcolo "M. Picone", Consiglio Nazionale delle Ricerche, Rome, Italy
e-mail: e.cristiani@iac.cnr.it

S. Tozza (✉)
Department of Mathematics, Alma Mater Studiorum Università di Bologna, Bologna, Italy
e-mail: silvia.tozza@unibo.it

E. Cristiani et al. (eds.), *Mathematical Methods for Objects Reconstruction*,
Springer INdAM Series 54, https://doi.org/10.1007/978-981-99-0776-2_1

Assuming the surface is represented by a graph $z = u(x, y)$, for every point (x, y) in the plane, that equation gives the link between the normal $\mathbf{N}$ to the surface, some parameters related to the light (that we simply denote by a vector $\mathbf{\Lambda}$) and the final image I. Note that this is a continuous representation of the image where every pixel is associated with a real value between 0 and 1 (the classical convention is to associate 0 to black and 1 to white). In practice, the points are discrete with integer coordinates and the values of I have discrete values between 0 and 255 for a gray-level image (the extension to color images is straightforward via the three channels RGB). Many authors have contributed to 3D reconstruction following the pioneering work by Horn [34, 35]. Many techniques have been proposed by researchers in computer science, engineering, and mathematics, the interested reader can find in the surveys [26, 27] a long list of references mainly focused on variational methods and nonlinear partial differential equations (PDEs).

In this chapter we focus on the problem to reconstruct the depth of a surface starting from one or more gray-value images, so that the datum is the shading contained in the input image(s). It should be mentioned that considering different input data, we can describe various problems belonging to the same class of Shape-from-X problems. These problems share the same goal (recover the 3D shape of the object(s)) and use a single viewpoint. Among these are the Shape-from-Polarization problem [78], the Shape-from-Template problem [13], the Shape-from-Texture problem [76], the Shape-from-Defocus problem [31], etc.

We concentrate our presentation on the methods based on first order nonlinear PDEs, and in particular eikonal type equations, since these equations also appear in the solution of some problems related to 3D printing, e.g., overhang construction and optimization. In this way we try to offer a unified approach that starts from one or more images, allows for the reconstruction of the surface, and finally brings to 3D printing of the solid object. Note that in this approach the theoretical framework is given by the theory of viscosity solutions for Hamilton–Jacobi equations, see [12] and the references therein.

The main focus here is on the modeling of some problems coming from computer vision and 3D printing in a unified framework, for this reason we just sketch some numerical approximation schemes. The interested reader can find in the books [30, 51, 61] many information on finite difference and semi-Lagrangian schemes for these problems.

The chapter is organized as follows: In Sect. 2, we present the standard approach for SfS based on a single image considering different models for the surface reflectance, the camera, and the light source. They all lead to a Hamilton–Jacobi equation complemented with various boundary conditions. Section 3 is devoted to the models based on more input images, the typical example comes either for a couple of images taken from different point of view (stereo vision) or from the same viewpoint but under different light source directions (photometric stereo). Since every image correspond to a different equation, we will have a system of equations similar to those examined in the previous section. We conclude the first part on 3D reconstruction with Sect. 4, which gives some hints on the numerical approximation of one of the setups illustrated in Sect. 2.

From Sect. 5 we deal with 3D printing. We begin recalling the level-set method for front propagation problems as a common tool for various problems related to 3D printing. In Sect. 6 we focus on the overhang issue, presenting three different methods for attacking this important problem. In Sect. 7 we address instead the problem of creating the infill structure of the object to be printed, also giving some practical examples.

We conclude the chapter with final remarks and two appendices, where we give some basic information about STL and G-code, two file formats commonly used in 3D printing.

2 Modeling with a Single Input Image

The modeling of the SfS problem depends on how we describe the three major elements that contribute to determine an image from the light reflection: the light characteristics, the reflectance properties of the surface and the camera. In fact, we can have various light sources as well as different reflectance properties of the surface and one or more cameras. So we start examining more in detail these three components.

2.1 *Modelization of the Surface Reflectance*

Depending on how we explicitly describe the function R in Eq. (1), different reflectance models arise. The simplest and more popular model in literature follows the Lambert's cosine law, which establishes that the intensity of a point in the image I in (1) is directly proportional to the cosine of the angle θ_i between the surface normal and the direction of the incident light. The Lambertian model can therefore be represented by the following equation

$$I(x, y) = \gamma_D \mathbf{N}(x, y) \cdot \mathbf{L} = \gamma_D \cos(\theta_i), \tag{2}$$

where $\mathbf{L}$ is a unit vector pointing towards the light source and γ_D denotes the diffuse albedo, i.e., the reflective power of a diffuse surface. This is a purely diffuse model, since the light is reflected by the surface in all directions, without taking into account the viewer direction, as visible in Fig. 1.

Another diffuse model is that proposed by Oren and Nayar in the 1990s [49, 50]. This model is suitable for rough surfaces, modelled as a set of facets with different slopes, each of them with Lambertian reflectance properties. The brightness equation for this model is

$$I(x, y) = \cos(\theta_i)(A + B \sin(\alpha) \tan(\beta) \max\{0, \cos(\varphi_r - \varphi_i)\}), \tag{3}$$

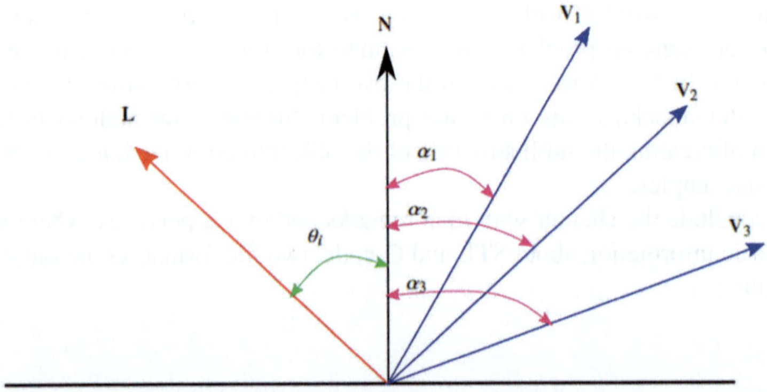

Fig. 1 Lambertian reflectance depends only on the incident angle θ_i. Three different viewers $\mathbf{V}_1$, $\mathbf{V}_2$, and $\mathbf{V}_3$ do not detect any difference in radiance

where θ_i denotes the incident angle, θ_r indicates the angle between the observer $\mathbf{V}$ and the normal $\mathbf{N}$ directions, α and β are defined as $\alpha = \max\{\theta_i, \theta_r\}$ and $\beta = \min\{\theta_i, \theta_r\}$. φ_i and φ_r are, respectively, the angles between the projection of the light source direction $\mathbf{L}$, or the projection of the viewer direction $\mathbf{V}$, and the x axis onto the (x, y)-plane. The constants A and B are nonnegative quantities and depend on the roughness parameter σ as follows

$$A = 1 - 0.5\,\sigma^2(\sigma^2 + 0.33)^{-1}$$
$$B = 0.45\sigma^2(\sigma^2 + 0.09)^{-1}.$$

More in general, we can define a reflectance model as sum of three components, which are the ambient, the diffuse, and the specular components (cf. [65]):

$$I(x, y) = k_A I_A(x, y) + k_D I_D(x, y) + k_S I_S(x, y), \tag{4}$$

with k_A, k_D, and k_S denoting the percentage of the three components, such that $k_A + k_D + k_S \le 1$ (if we do not consider absorption phenomena, this sum should be equal to one).

Following this setting, we can define, for example, the Phong model [53] as

$$I(x, y) = k_A I_A(x, y) + k_D \gamma_D(x, y)(\cos\theta_i) + k_S \gamma_S(x, y)(\cos\theta_s)^{\alpha}, \tag{5}$$

where the diffuse model is defined as in the Lambertian model, and the third term describes the specular light component as a power of the cosine of the angle θ_s between the unit vectors $\mathbf{V}$ and $\mathbf{R}(x, y)$ with $\mathbf{R}$ indicating the reflection of the light $\mathbf{L}$ on the surface, α represents the characteristics of specular reflection of a material, and $\gamma_S(x, y)$ denotes the specular albedo.

In 1977 Blinn [14] proposed a modification of the Phong model via an intermediate vector **H**, which bisects the angle between the unit vectors of the viewer **V** and the light **L**. The brightness equation in this case is

$$I(x, y) = k_A I_A(x, y) + k_D \gamma_D(x, y)(\cos\theta_i) + k_S \gamma_S(x, y)(\cos\delta)^c, \tag{6}$$

where δ is the angle between **H** and the unit normal **N**, and c measures the shininess of the surface.

Several other reflectance models can be mentioned: the Torrance-Sparrow model [63], simplified by Healey and Binford [33], the Cook-Torrance model [21], the Wolff diffuse reflection model for smooth surfaces [77], the Wolff-Oren-Nayar model that works for smooth and rough surfaces [79], the Ward model for surfaces with hybrid reflection [74], etc.

2.2 *Some Hints on Theoretical Issues: Viscosity Solutions and Boundary Conditions*

In the presentation of the previous models we have always assumed that the surface u is sufficiently regular so that the normal to the surface is always defined. Clearly, this is a very restrictive assumption that is not satisfied for real objects represented in our images. A more reasonable assumption would require almost everywhere (a.e.) regularity, as for a pyramid. This will guarantee the uniqueness of the normal vector everywhere but some curves corresponding to the singular points of the surface, i.e., these points form the edges of the object. One could accept solutions in an a.e. sense but this is too much since this simple extension can easily produce infinitely many solutions to the same equation even fixing boundary conditions, this is well known even for the simple eikonal equation on an interval

$$|\nabla u(x)| = f(x) \qquad x \in (a, b) \tag{7}$$

with homogeneous boundary conditions

$$u(a) = u(b) = 0 \tag{8}$$

whenever f vanishes at one point. One way to select a unique solution is to adopt the definition of viscosity solution v introduced by Crandall and Lions in the 80s. We will not give this definition here, but it can be easily found, e.g., in [12]. For the sequel, just keep in mind that a viscosity solution is an a.e. solution with some additional properties, such as being the limit for ε going to 0 of the solution v^ε of a second order problem

$$-\varepsilon u_{xx} + |\nabla u(x)| = f(x) \qquad x \in (a, b) \tag{9}$$

with homogeneous boundary conditions

$$u(a) = u(b) = 0. \tag{10}$$

In some cases, the classical definition is not enough to select a unique solution and additional properties have been considered, e.g., the fact that v must be also the maximal solution, i.e., $v \geq w$ for every w that is an a.e. solution.

Boundary Conditions
Another crucial point in the theory of viscosity solutions is the treatment of boundary conditions. All the classical boundary conditions can be considered, but they must be understood (and implemented) in a weak sense.

For the SfS problem, we will consider either Dirichlet boundary conditions

$$u(x) = g(x) \qquad x \in \Gamma = \partial\Omega \tag{11}$$

where $g = 0$ means that the object is standing on a flat background and a generic g means that we know the height of the surface on Γ (this is the case for rotational surfaces).

For the front propagation problem, the typical boundary condition will be the homogeneous Neumann boundary condition

$$\frac{\partial u(x)}{\partial \mathbf{N}} = 0 \qquad x \in \Gamma = \partial\Omega. \tag{12}$$

We refer again to [12] for a detailed analysis of these problems and to [30, 51, 61] for their numerical approximation.

2.3 Modelization of the Camera and the Light

Assuming that the light source is located at infinity, all light vectors are parallel and we can represent the light direction by a constant vector $\mathbf{L} = (l_1, l_2, l_3)$. We assume that the light source is above the surface to be reconstructed, so $l_3 > 0$.

If we assume that the image is obtained by orthographic projection of the scene, then we can define the surface $\mathcal{S}$ as

$$\mathcal{S} = \{(x, y, u(x, y)) : (x, y) \in \overline{\Omega}\}, \tag{13}$$

where Ω denotes the reconstruction domain. Assuming that the surface is regular, a normal vector is defined by

$$\mathbf{n}(x, y) = (-\nabla u(x, y), 1). \tag{14}$$

Note that orthogonal projection model is rather restrictive since it requires that the distance between the camera and the object is larger than 5–6 times the dimension of the object; however, it is enough to explain the main features of the SfS problem. As an example, the Lambertian model (2) under orthographic projection with a single light source located at infinity and uniform albedo constantly equal to one, $\gamma_D \equiv 1$, which means that all the incident light is reflected, leads to the following partial differential equation (PDE) in the unknown u

$$I(x, y) = \frac{-\nabla u(x, y) \cdot (l_1, l_2) + l_3}{\sqrt{1 + |\nabla u(x, y)|^2}}. \tag{15}$$

For the SfS problem, we look for a function $u : \overline{\Omega} \to \mathbb{R}$ satisfying the Eq. (15) $\forall (x, y) \in \Omega$, given an image $I(x, y)$ and a light source direction $\mathbf{L}$.

In the particular case of a vertical light source, i.e., $\mathbf{L} = (0, 0, 1)$, the PDE in (15) reduces to the so-called *Eikonal equation*

$$|\nabla u(x, y)| = \sqrt{\frac{1}{I(x, y)^2} - 1}\,. \tag{16}$$

Assuming that the image is obtained by a perspective projection of the scene, we need to distinguish different cases depending on the position of the light source.

Supposing a light source located at infinity, the scene can be represented by the following surface

$$\mathcal{S} = \{u(x, y)(x, y, -f) : (x, y) \in \overline{\Omega}\}, \tag{17}$$

where $f \geq 0$ denotes the focal length, i.e., the distance between the retinal plane and the optical center. This is the case of a "pinhole" camera, that is a simple camera without a lens but with a tiny aperture (the so-called pinhole). In this case, the unit normal vector will be

$$\mathbf{N}(x, y) = \frac{\mathbf{n}(x, y)}{|\mathbf{n}(x, y)|} = \frac{(f \nabla u(x, y), u(x, y) + (x, y) \cdot \nabla u(x, y))}{\sqrt{f^2 |\nabla u(x, y)|^2 + (u(x, y) + (x, y) \cdot \nabla u(x, y))^2}} \tag{18}$$

and so the Lambertian model (2) with uniform albedo $\gamma_D \equiv 1$ in this case will lead to the following PDE

$$I(x, y) = \frac{f\ (l_1, l_2) \cdot \nabla u(x, y) + l_3(u(x, y) + (x, y) \cdot \nabla u(x, y))}{\sqrt{f^2 |\nabla u(x, y)|^2 + (u(x, y) + (x, y) \cdot \nabla u(x, y))^2}}. \tag{19}$$

Note that assuming the surface visible, i.e., in front of the retinal plane, then the function $u(x, y) \geq 1, \forall (x, y) \in \overline{\Omega}$ [55]. Hence, operating the change of variable $v(x, y) = \ln u(x, y)$, the new variable assumes values greater or equal to 0, $v \geq 0$.

The new PDE for the variable v of Eq. (19) becomes

$$I(x, y) = \frac{(f\ (l_1, l_2) + l_3(x, y)) \cdot \nabla v(x, y) + l_3}{\sqrt{f^2|\nabla v(x, y)|^2 + (1 + (x, y) \cdot \nabla v(x, y))^2}}. \tag{20}$$

Now, let us still consider a perspective projection, but with a single point light source located at the optical center of the camera. This case corresponds to a camera equipped with a flash in a dark place. In this case the parametrization of the scene is a little bit different from the previous case, with a surface $\mathcal{S}$ defined as

$$\mathcal{S} = \left\{ \frac{f u(x, y)}{\sqrt{f^2 + |(x, y)|^2}}(x, y, -f) : (x, y) \in \overline{\Omega} \right\}. \tag{21}$$

For such a surface, the unit normal vector is

$$\mathbf{N}(x, y) = \frac{(f\ \nabla u(x, y) - \frac{f u(x,y)}{|(x,y)|^2 + f^2}\ (x, y), \nabla u(x, y) \cdot (x, y) + \frac{f u(x,y)}{|(x,y)|^2 + f^2}\ f)}{\sqrt{f^2|\nabla u(x, y)|^2 + (\nabla u(x, y) \cdot (x, y))^2 + u(x, y)^2 \frac{f^2}{(f^2 + |(x,y)|^2)}}} \tag{22}$$

and the unit light vector depends in this case on the point (x, y) of $\mathcal{S}$, that is

$$\mathbf{L}(\mathcal{S}(x, y)) = \frac{(-(x, y), f)}{\sqrt{|(x, y)|^2 + f^2}}.$$

Under this setup, the Lambertian model (2) with uniform albedo $\gamma_D \equiv 1$ leads to the following PDE

$$I(x, y)\sqrt{\left(\frac{|(x, y)|^2 + f^2}{f^2}\right)\left[f^2|\nabla u(x, y)|^2 + ((x, y) \cdot \nabla u(x, y))^2\right] + u(x, y)^2} - u(x, y) = 0. \tag{23}$$

As done before, assuming the surface visible, then

$$u(x, y) \geq \frac{|(x, y)|^2 + f^2}{f^2} \geq 1, \forall (x, y) \in \overline{\Omega}.$$

By the same change of variable $v(x, y) = \ln u(x, y)$, considering Lambertian reflectance in this setup we arrive to the following Hamilton–Jacobi (HJ) equation

$$I(x, y)\sqrt{\left(\frac{|(x, y)|^2 + f^2}{f^2}\right)\left[f^2|\nabla v(x, y)|^2 + ((x, y) \cdot \nabla v(x, y))^2\right] + 1} - 1 = 0. \tag{24}$$

Regarding to the last setup (perspective projection with a light source located at the optical center of the camera), some researchers have introduced a light attenuation term in the brightness equation related to the model considered for real-world experiments (see, e.g., [1, 2, 56, 57]), and in [18] the well-posedness of the perspective Shape-from-Shading problem for several non-Lambertian models in the context of viscosity solutions is provided, thanks to the introduction of this term. This factor takes into account the distance between the surface and the light source and, together with the use of non-Lambertian reflectance models, seems to be useful for better describing the real images.

3 Modeling with More Input Images

Unfortunately, the classical SfS with a single input image is, in general, an ill-posed problem in both the classical and the weak sense, due to the well-known concave/convex ambiguity (see the surveys [26, 82]). It is possible to overcome this issue by adding information, as, e.g., setting the value of the height u at each point of maximum brightness [43] or choosing as solution the maximal one, which is proven to be unique [17]. Under an orthographic projection, the ill-posedness is still present even if we consider non-Lambertian reflectance models [65]. Using a perspective projection, in some particular cases an ambiguity is still visible [15, 64].

A natural way to add information is considering more than one input image. We will explain in the following two different approaches which allow to get well-posedness.

3.1 Photometric Stereo Technique

The Photometric Stereo SfS (PS-SfS) problem considers more than one input image, taken from the same point of view but with a different light source for each image (see [27] for a comprehensive introduction on this problem). Just to fix the idea, let us consider two pictures of the same surface, whose intensities are I_1, I_2, respectively, taken from the light sources $\mathbf{L}'$, $\mathbf{L}''$. From the mathematical viewpoint, assuming an orthographic projection of the scene, Lambertian reflectance model, and light sources located at infinity in the direction of the two versors $\mathbf{L}'$, $\mathbf{L}''$, this means considering a system of two equations as

$$\begin{aligned} I_1(x, y) &= \gamma_D \mathbf{N}(x, y) \cdot \mathbf{L}' , \\ I_2(x, y) &= \gamma_D \mathbf{N}(x, y) \cdot \mathbf{L}'' . \end{aligned} \tag{25}$$

Following a differential approach, writing $\mathbf{N}(x, y) = \dfrac{\mathbf{n}(x, y)}{|\mathbf{n}(x, y)|}$, with $\mathbf{n}(x, y)$ defined as in (14), this corresponds to a system of nonlinear PDEs, in which the nonlinear term is common to both equations and comes from the normalization of the normal vector. Isolating this nonlinear term and the diffuse albedo, the system (25) becomes the following hyperbolic linear problem

$$\begin{cases} \mathbf{b}(x, y) \cdot \nabla u(x, y) = f(x, y), & \text{a.e. } (x, y) \in \Omega, \\ u(x, y) = g(x, y), & \forall (x, y) \in \partial\Omega, \end{cases} \tag{26}$$

to which we added Dirichlet boundary conditions, and where

$$\mathbf{b}(x, y) = (I_2(x, y)l_1' - I_1(x, y)l_1'', I_2(x, y)l_2' - I_1(x, y)l_2'') \tag{27}$$

and

$$f(x, y) = I_2(x, y)l_3' - I_1(x, y)l_3''. \tag{28}$$

It can be shown that the problem (26) is well-posed [45], getting over the concave/convex ambiguity, and it is also albedo independent, a powerful property for real applications. Since in real application is difficult to know the correct boundary conditions, in [46] the authors shown that it is possible to reconstruct uniquely the surface starting from three images taken from three non-coplanar light sources. Knowing a priori symmetric properties of the surface, the number of required images reduces to a single image generated from any not vertical light source for surfaces with four symmetry straight lines [46]. Non-Lambertian reflectance models have been considered for the PS-SfS problem (see, e.g., [20, 32, 83]), and well-posedness has been proved also including specular effects, which allow to reduce artifacts [67].

The PS-SfS problem, as well as the problems illustrated in the previous section, can also be solved via a non-differential approach, in which the unknown is the outgoing normal vector to the surface, defined in (14). Differently from the previous differential approach, which works globally, this second approach is a local one, that works pixel by pixel in order to find the normal vector field of the unknown surface and then it reconstructs the surface $z = u(x, y)$ all over the domain using the gradient field [24, 25]. Integrability constraint must be imposed in order to limit the ambiguities [48, 81]. We refer the interested reader to [58, 59] for a survey on normal integration methods.

3.2 Multi-View SfS

The Multi-View SfS (MV-SfS) problem considers more than one input image taken from several points of view, but under the same light source for all the images. Differently from the PS-SfS problem, here we have to solve the "matching issue" (also known as *correspondence problem*), since considering a different viewpoint for each image, the same part of the object(s) we want to reconstruct is located in different pixels of the other images.

To fix the ideas, you can think to the human binocular vision so the relative depth information should be extracted by examining the relative positions of object(s) in the two images. Under suitable assumptions, it is possible to achieve a well-posed problem. As an example, we can follow the Chambolle's approach in [19], which gets the well-posedness in the case of two cameras under the following assumptions:

A1 smooth surface ($u \in C^1$)
A2 no shadows on the surface
A3 noise-free images
A4 single light source $\mathbf{L}$ located at infinity
A5 Lambertian reflectance model
A6 pictures taken with a set of parallel "standard stereo cameras" (i.e., two ideal cameras, in which the image is obtained by a simple projection through a pointwise lens, whose coordinate systems are parallel to each other and whose focal distance f is the same)
A7 no occluded zones on the two images
A8 surface represented by a function $Z = u(X, Y)$ where the plane (X, Y) is parallel to the focal planes of the cameras and Z remains bounded ($Z \leq C$) The two lens are in $(t/2, B, C)$ and $(-t/2, B, C)$, where $t > 0$ denotes the translation parameter in the horizontal axis. A point in the scene (X, Y, Z), with $Z \leq C$, appears on the images I_1 and I_2 in (x_1, y) and (x_2, y) where

$$x_1 = f\frac{X - t/2}{C - Z}, \quad x_2 = f\frac{X + t/2}{C - Z}, \quad y = f\frac{Y - B}{C - Z}. \tag{29}$$

Let Ω_1 be the set of all the points of the first image that appear also on the second one and let us assume that it is a connected set. It should be possible to reconstruct the surface corresponding to the set Ω_1 without ambiguity, if the direction of the light source $\mathbf{L}$ is not too close to the direction (X, Y) of the focal planes, and if the disparity function, which is given by $d := f\frac{t}{u}$, where t is the baseline length, is already known on the boundary of the set Ω_1.

In one dimension, the problem amounts to look for a disparity function $d(x_1) : \Omega_1 \to \mathbb{R}$ verifying the matching condition

$$\forall x_1 \in \Omega_1, \qquad I_1(x_1) = I_2(x_1 + d(x_1)). \tag{30}$$

Let us define the set

$$A = \{a \in \Omega_1 | I_1 \text{ is constant in a neighborhood of } a\}.$$

Chambolle has proved [19, Theorem 1] that Eq. (30), associated with the knowledge of Ω_1 and the value of d at one point $a_0 \in \overline{\Omega_1}$, is enough to recover the disparity everywhere where I_1 is not locally constant. Note that Eq. (30) gives no information on the value of the disparity in A.

Once Ω_1 and the disparity d are known, the part of the surface corresponding to the set Ω_1 can be recovered inverting the formulas in (29), obtaining a parametrized surface and then a normal vector to the surface according to Assumption A1. By the change of variable $v = \log |d|$, and taking into account Assumption A5, in 1D the SfS equation is

$$I_1 = \mathbf{L} \cdot \mathbf{N}, \tag{31}$$

where $\mathbf{L} = (l_1, l_2)$, with $l_2 < 0$, $\mathbf{N} = (N_1, N_2) = \frac{\mathbf{n}}{|\mathbf{n}|}$. By Assumption A8, we need to have $N_2 < 0$ and thus $N_2 = -\sqrt{1 - N_1^2}$, so that

$$\mathbf{N} = (N_1, N_2) = \left(l_1 I_1 \pm l_2 \sqrt{1 - I_1^2}, -\sqrt{1 - N_1^2} \right). \tag{32}$$

Hence, there are two possible values for $\mathbf{N}$, except when $I_1 = 1$. Thanks to Eq. (30), the disparity d is uniquely determined in $\Omega_1 \setminus A$. In $1D$, A is union of open segments $K = (x_0, x_1)$, on which I_1 is constant and (30) guarantees the uniqueness of $d(x_0)$ and $d(x_1)$, whereas (32) gives two possible values for the derivative $N'(x)$ on K. $\forall x \in K$, if the two values not coincide, only one of the two N' is compatible with the known values $d(x_0)$ and $d(x_1)$.

To ensure that d can be reconstructed in any case without ambiguity on the whole set Ω_1, we have to know, as a boundary condition, its value on $\partial\Omega_1 = \{\inf(\Omega_1), \sup(\Omega_1)\}$.

For the extension to the 2D case, we refer the interested reader to the Chambolle's paper [19], and the results on SfS based on viscosity solutions used therein for the uniqueness result [22, 42, 43].

4 An Example of Numerical Resolution

In this section, we intend to describe a case of numerical resolution for the classical SfS, whose ingredients that play a role for the 3D reconstruction have been illustrated in Sect. 2. As an example, let us fix an orthographic projection, so that the surface normal vector is defined as in Eq. (14), and a single light source located at infinity. Let us consider three different reflectance models: the Lambertian model, the Oren-Nayar model, and the Phong one, whose brightness equations are reported in Eqs. (2), (3), and (5), respectively. As explained in [65, 66], it is possible to write the three different formulations of these models in a unified fixed-point problem as follows:

$$\mu v(x, y) = T^M(x, y, v, \nabla v), \quad \text{for } (x, y) \in \Omega, \tag{33}$$

where M denotes the acronyms of the three models, i.e., $M = L, ON, PH$, v is the new variable introduced by the exponential Kružkov change of variable [39] $\mu v(x, y) = 1 - e^{-\mu u(x,y)}$, which is useful for analytical and numerical reasons. In fact, setting the problem in the new variable, v will have values in $[0, 1/\mu]$ instead of $[0, \infty)$ as the original variable u so an upper bound will be easy to find. The parameter μ is a free constant which does not have a specific physical meaning in the SfS problem, but it can play an important role also in the convergence proof (see the remark following the end of Theorem 1 in [65]). The operator T^M can be defined as

$$T^M := \min_{a \in \partial B_3} \{\mathbf{b}^M(x, y, a) \cdot \nabla v(x, y) + f^M(x, y, z, a, v(x, y))\}, \tag{34}$$

where the vector field $\mathbf{b}^M$ and the cost f^M vary according to the specific model and to the case.

This unified formulation gives the big advantage to solve numerically different problems in a unified way. In order to obtain the fully discrete approximation, we adopt here the semi-Lagrangian approach described in the book [30], so that we get

$$W_i = \widehat{T}_i^M(\mathbf{W}), \tag{35}$$

where $\mathbf{W}$ is the vector solution giving the approximation of the height u at every node x_i of the grid, and we use the notation $W_i = w(x_i)$, i indicating a multi-index, $i = (i_1, i_2)$. Denoting by G the global number of nodes in the grid, the operator corresponding to a general oblique light source is $\widehat{T}^M : \mathbb{R}^G \to \mathbb{R}^G$ that is defined componentwise by

$$\widehat{T}_i^M(\mathbf{W}) := \min_{a \in \partial B_3} \left\{ e^{-\mu h} I[\mathbf{W}]\left(x_i^+\right) - \tau F^M(x_i, z, a) \right\} + \tau, \tag{36}$$

where $I[\mathbf{W}]$ represents an interpolation operator based on the values at the grid nodes, and

$$x_i^+ := x_i + h b^M(x_i, a) \tag{37}$$

$$\tau := (1 - e^{-\mu h})/\mu \tag{38}$$

$$F^M(x_i, z, a) := P^M(x_i, z) a_3 (1 - \mu W_i) \tag{39}$$

$$P^M : \Omega \times \mathbb{R} \to \mathbb{R} \text{ is continuous and nonnegative.} \tag{40}$$

Since $w(x_i + h b^M(x_i, a))$ is approximated via $I[\mathbf{W}]$ by interpolation on $\mathbf{W}$ (which is defined on the grid G), it is important to use a monotone interpolation in order to preserve the properties of the continuous operator T^M in the discretization. To this end, the typical choice is to apply a piecewise linear (or bilinear) interpolation operator $I_1[\mathbf{W}] : \Omega \to \mathbb{R}$ which allows to define a function defined for every $x \in \Omega$ (and not only on the nodes)

$$w(x) = I_1[\mathbf{W}](x) = \sum_j \lambda_{ij}(a) W_j, \tag{41}$$

where

$$\sum_j \lambda_{ij}(a) = 1 \quad \text{for} \quad x = \sum_j \lambda_{ij}(a) x_j. \tag{42}$$

A simple explanation for (41)–(42) is that the coefficients $\lambda_{ij}(a)$ represent the local coordinates of the point x with respect to the grid nodes (see [30] for more details and other choices of interpolation operators). Clearly, in (36) we apply the interpolation operator to the point $x_i^+ = x_i + hb^M(x_i, a)$ and we denote by w the function defined by $I_1[\mathbf{W}]$.

Under suitable assumptions, it is possible to prove that the fixed-point operator $\widehat{T}^M$ is a contraction mapping in $L^\infty([0, 1/\mu)^G)$, is monotone, and $0 \le \mathbf{W} \le \frac{1}{\mu}$ implies $0 \le \widehat{T}^M(\mathbf{W}) \le \frac{1}{\mu}$. These properties allow to prove a convergence result for the algorithm based on the fixed-point iteration

$$\begin{cases} \mathbf{W}^n = \widehat{T}^M(\mathbf{W}^{n-1}), \\ \mathbf{W}^0 \text{ given.} \end{cases} \tag{43}$$

We refer the interested reader to [65] for the proofs and more details. An example of numerical results by using the semi-Lagrangian scheme illustrated above is visible in Fig. 3, where the 3D reconstructions obtained by using the operators related to the three reflectance models are shown. This figure is related to a real image of chess horse used as input image and visible in Fig. 2 with the corresponding mask adopted. For this numerical test, the Phong model performs clearly better, being the surface shiny.

Fig. 2 From left to right: Input image and related mask adopted for the black chess horse test. Pictures taken and adapted from [66]

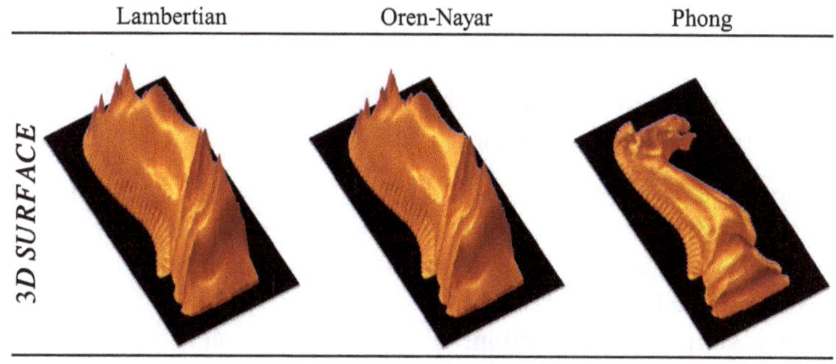

Fig. 3 Black chess horse: 3D reconstructions related to the three models with $\sigma = 0.2$ in the Oren-Nayar model, and $k_S = 0.8$, $\alpha = 1$ for the Phong model. Light source $\mathbf{L} = (0, 0, 1)$, viewer $\mathbf{V} = (0, 0, 1)$. Pictures taken and adapted from [66]

A Funny Counterexample of Non-uniqueness

Just for fun, we have tried to 3D-print the surface reconstructed from the famous Lena image under the assumptions of orthographic Lambertian model: although the surface mismatches completely the real person, a picture taken from above correctly returns the input image, as expected!

5 Moving from 3D Vision to 3D Printing

Surfaces like the ones depicted in Fig. 3, which represent the graphs of different functions, can be easily transformed in watertight (closed) surfaces of 3D objects, in order to manufacture them with suitable 3D printers. However, when it comes the time to actually 3D-print the object, possibly reconstructed via SfS techniques, many additional issues could arise. In the following we will review some of them, with special emphasis to those which are most related to 3D vision and/or Hamilton–Jacobi equations, level-set method and front propagation problems [51, 61]. The related basic mathematical theory will be recalled in Sect. 5.2.

5.1 Overview

Here we list a brief summary of problems often encountered in 3D printing. For broader overviews, we refer the reader to [28, 44, 62] and the references therein.

- **Partitioning**. Sometimes it is necessary to divide a 3D model into multiple printable pieces, so as to save the space, to reduce the printing time, or to make a large model printable by small printers [11]. This problem was attacked by means of a level-set based approach in [80].
- **Overhang, supports, orientation**. 3D printers based on fused deposition modeling (FDM) technology create solid object layer by layer, starting from the lowest one. As a consequence, each layer can only be deposited on top of an existing surface, otherwise the print material falls down and solidifies in a disorderly manner. Little exceptions can be actually handled, i.e., the upper layer can protrude over the lower layer within a certain limit angle $\bar{\alpha}$. Typically $\bar{\alpha} = \frac{\pi}{4}$, because of the so-called 45 degree rule, though it actually depends on the 3D printer settings, print material, cooling, etc. In recent years many solutions were proposed to deal with this problem: first of all, the object can be rotated to find an orientation, if any, which does not show overhangs [29]. Another widely adopted solution consists in adding *scaffolds* during construction, in order to support the overhanging regions. In this respect, it is important to note that scaffolds represent an additional source of printed material and printing time and have to be manually removed at the end of the process. This is cumbersome, time-consuming and reduces the quality of the final product, see, e.g., [16] and the references therein. Finally, overhangs issues can be fixed *in the design phase of the object*, running a shape optimization algorithm which modifies the unprintable object in order to make it printable while preserving specific features [4, 7–9, 41].
  In the following sections we recall three approaches based on nonlinear PDEs. All of them recast the problem of overhangs in a front propagation problem and then use the level-set method to solve the problem.
- **Infill.** Printing fully solid objects is often not convenient because of the large quantity of material to be used. The problem reduces to finding the optimal inner structure for saving material while keeping the desired rigidity and printable features. In Sect. 7 we will detail a method inspired by Dhanik and Xirouchakis [23], and Kimmel and Bruckstein [38] and based on the front propagation problem to compute ad hoc infill patterns.
- **Appearance.** A problem about 3D printing which is strongly related to 3D vision and Shape-from-Shading problem regards the way the objects appear, i.e., how they reflect light when illuminated. This problem is very important when the goal is to replicate (multi-material) real objects with complex reflectance features using a single printing material, possibly with simple Lambertian reflectance. The idea is to print the object with small-scale ripples on the surface in order to control the reflectance properties for a given direction (or all directions) of the light source. For example, the approach proposed in [40] allows one to

create a shape which reproduces the appearance of costly materials (e.g., velvet) using cheaper ones, duly orienting small facets on the surface and then covering them by ad hoc varnish. In this way it is possible to reproduce a given target Bidirectional Reflectance Distribution Function (BRDF) [47]. Note that this is the same ingredient used in Shape-from-Shading for the opposite problem, i.e., reconstruct the shape from reflectance features. The possibility of controlling the reflectance properties of an object by means of its physical shape opens many interesting possibilities, one can refer to [54, 60, 75] for some examples.

- **Slicing and toolpath generation.** Once the object to be printed is defined, typically in terms of triangles covering its surface, one has to cut it with a sequence of parallel planes in order to compute the *layers* used in the printing process. This procedure is not trivial in case of very complicated (self-intersecting, overlapped) objects. Once the layers are created, the exact trajectory of the nozzle must be defined in order to cover all the layer points; see, among others, [36, 37]. Interestingly, this problem can be seen as a generalized traveling salesman problem.

In the following we will review some mathematical approaches for the overhang and the infill problems, all based on the level-set method for front propagation problems. Before that, we recall some basic mathematical results for the reader's convenience.

5.2 *Front Propagation Problem, Level-set Method and the Eikonal Equation*

The level-set method was introduced in [52] to track the evolution of a $(d-1)$-dimensional surface (the front) embedded in $\mathbb{R}^d$ evolving according to a given velocity vector field $\mathbf{v} : \mathbb{R}^+ \times \mathbb{R}^d \to \mathbb{R}^d$. Let us briefly recall the method in the case of $d = 3$. It is given a bounded closed surface $\Sigma_0 : U \subset \mathbb{R}^2 \to \mathbb{R}^3$ which represents the front at initial time $t = 0$. We denote by Σ_t its (unknown) evolution under the action of $\mathbf{v}$ at time t and by Ω_t the 3D domain strictly contained in Σ_t so that $\Sigma_t = \partial\Omega_t$, for all $t \geq 0$.

The main idea of the level-set method consists in increasing the space dimension of the problem by 1, in order to represent the 3D surface as a level-set of a 4D function $\varphi(t, x, y, z) : \mathbb{R}^+ \times \mathbb{R}^3 \to \mathbb{R}$. This function is defined in such a way that

$$\Sigma_t = \{(x, y, z) \; : \; \varphi(t, x, y, z) = 0\}, \quad \forall t \geq 0. \tag{44}$$

In this way the surface is recovered as the zero level-set of φ at any time. As initial condition for φ, one chooses a function which changes sign at the interface,

$$\varphi(0, x, y, z) \begin{cases} > 0, \text{ if } (x, y, z) \notin \overline{\Omega_0}, \\ = 0, \text{ if } (x, y, z) \in \Sigma_0, \\ < 0, \text{ if } (x, y, z) \in \Omega_0. \end{cases} \tag{45}$$

A typical choice for $\varphi(0, x, y, z)$ is the signed distance function from Σ_0, although is it not smooth and can lead to numerical issues. It can be proved [61] that the level-set function φ at any later time satisfies the following Hamilton–Jacobi equation

$$\partial_t \varphi(t, x, y, z) + \mathbf{v}(t, x, y, z) \cdot \nabla \varphi(t, x, y, z) = 0, \qquad t \in \mathbb{R}^+, \ (x, y, z) \in \mathbb{R}^3, \tag{46}$$

with an initial condition $\varphi(0, x, y, z) = \varphi_0(x, y, z)$ satisfying (45). Here $\nabla = (\partial_x, \partial_y, \partial_z)$ denotes the gradient with respect to the space variables only.

Several geometrical properties of the evolving surface can be still described in the new setting, again by means of its level-set function φ. For example, it is possible to write the unit exterior normal $\mathbf{N}$ and the (mean) curvature κ in terms of φ and its derivatives. More precisely, we have

$$\mathbf{N} = \frac{\nabla \varphi}{|\nabla \varphi|} \quad \text{and} \quad \kappa = \nabla \cdot \mathbf{N}. \tag{47}$$

If the evolution of the surface occurs in the normal direction of the surface itself, i.e., the vector field has the form $\mathbf{v} = v\mathbf{N}$ for some scalar function $v : \mathbb{R}^+ \times \mathbb{R}^3 \to \mathbb{R}$, the Eq. (46) turns into the evolutive eikonal equation

$$\partial_t \varphi(t, x, y, z) + v(t, x, y, z)|\nabla \varphi(t, x, y, z)| = 0. \tag{48}$$

Moreover, if the scalar velocity field v also depends on the direction of propagation $\mathbf{N}$, using (47) we get the *anisotropic eikonal equation*

$$\partial_t \varphi(t, x, y, z) + v\left(t, x, y, z, \frac{\nabla \varphi}{|\nabla \varphi|}\right) |\nabla \varphi(t, x, y, z)| = 0. \tag{49}$$

Finally, in the particular case of a constant-sign time-independent velocity field $v(x, y, z, \mathbf{N}) > 0$ or $v(x, y, z, \mathbf{N}) < 0$ for all (x, y, z) we are guaranteed that the surface evolution is *monotone*, i.e., the surface is either enlarging or shrinking for all times t. In this case, Eq. (49) can be written in the following stationary form

$$v\left(x, y, z, \frac{\nabla T}{|\nabla T|}\right) |\nabla T(x, y, z)| = 1 \tag{50}$$

and the surface Σ can be recovered from T as

$$\Sigma_t = \{(x, y, z) \ : \ T(x, y, z) = t\}, \quad \forall t \geq 0.$$

All these equations are particular Hamilton–Jacobi equations for which many theoretical results and numerical methods were developed in the last years. We refer the reader to [30] for details.

5.3 Computation of the Signed Distance Function from a Surface

The computation of the signed distance function φ_0 from Σ_0 is a problem *per se*. Speaking of 3D objects to be 3D-printed, we can assume that the surface Σ_0 is watertight and that it is given by means of a triangulation (typically in the form of a .STL file, see Appendix A). Each triangle (facet) f is characterized by the 3D coordinates of its three vertices. Moreover, vertices are oriented in order to distinguish the internal and the external side of the facet (right-hand rule).

Given a point $(x, y, z) \in \mathbb{R}^3$, it is easy to find the distance $d\big((x, y, z), f\big)$ between the point and the facet, so that the unsigned distance from the surface is simply given by

$$d((x, y, z), \Sigma_0) = \min_f d((x, y, z), f).$$

The computation of the distance's sign is instead more tricky since one has to check if the point is internal or external to the surface. One method to solve the problem relies on the fact that the solid angle subtended by the whole surface at a given point is maximal and equal to 4π if and only if the point is internal. Therefore, one can sum all the solid angles subtended by the facets at the point and check if it equals 4π or not. In the first case the point is internal to the surface, in the second case it is external.

Note that the solid angle itself should be *signed*, in the sense that it must be positive if the point looks at the internal part of the facet, negative otherwise. A nice algorithm to compute the signed solid angle between a point and a triangle was given by van Oosterom and Strackee [73].

6 Handling Overhangs

In this section we deal with the problem of overhangs, see Sect. 5.1. We recall three approaches recently proposed in the context of Applied Mathematics. All of them recast the problem of detecting/resolving overhangs in a *front propagation problem*.

6.1 Detecting Overhangs via Front Propagation

Here we recall the method proposed by van de Ven et al. in [68] and then extended by the same authors in [69–71]. The authors start noting the resemblance between the front propagation problem, see Sect. 5.2, and the additive manufacturing process where with every added layer, the lower boundary of the product advances. Therefore, the first layer, i.e., the layer leaning on the build plate, is considered as the initial front and it is propagated by solving the anisotropic stationary eikonal Eq. (50).

Overhangs are detected by solving the equation with two different velocity fields. In the first case, the velocity field is chosen is such a way that the front reaches, at the same time, all points of the object which are at the same height. In this case it is not actually needed to solve the equation since the arrival time is a measure of distance to the base plate, the minimum arrival time is simply given by

$$T_1(x, y, z) = \frac{(x, y, z) \cdot \mathbf{h}}{v_0}, \tag{51}$$

where $\mathbf{h}$ is a unit vector pointing in the build direction (assuming that the origin of the coordinate system is on the build plate), and v_0 is the propagation speed, which can be interpreted as the printing rate.

In the second case, which is the crucial one, the velocity field is chosen to be constant and equal to v_0 (thus recovering the classical isotropic eikonal equation) except when the front travels in a direction lower than the minimum allowable overhang angle. When this happens, the velocity is *decreased*, so that the arrival time will be larger than T_1 at the same point. In [70] this effect is obtained defining

$$v(\mathbf{a}; \bar{\alpha}) = \frac{v_0}{\max\{\tan(\bar{\alpha})\|\mathbf{Pa}\|, |\mathbf{h} \cdot \mathbf{a}|\}}, \tag{52}$$

where $\mathbf{a} = \frac{\nabla T}{|\nabla T|}$ is the direction of propagation, $\bar{\alpha}$ is a given overhang angle (parameter), and $\mathbf{P}$ is the projection on the plane orthogonal to $\mathbf{h}$, defined as $\mathbf{P} = \mathbf{I} - \mathbf{h} \otimes \mathbf{h}$, with $\otimes$ denoting the outer product. See [70] for a detailed motivation of this choice.

Finally, a straightforward comparison of the two solutions reveals the overhanging regions, associated with different arrival times of the front.

6.2 *Fixing Overhangs via Level-set Method 1: A Direct Approach*

Here we recall the method proposed by Cacace et al. in [16]. The idea is to recast the problem in the level-set framework considering the surface Σ of the object Ω to be 3D-printed as an evolving front. If the object is unprintable without scaffolds, its shape is modified letting its surface evolve under an *ad hoc* vector field $\mathbf{v}$ to be suitably defined, until it becomes fully printable, i.e., it has no hanging parts exceeding the limit angle α. The new object Ω_* is then actually printed and the difference $\Omega_* \backslash \Omega_0$ is finally removed. Note that the difference $\Omega_* \backslash \Omega_0$ can be easily identified by standard techniques and consequently printed with a different material (e.g., a soluble filament) or with a different printing resolution.

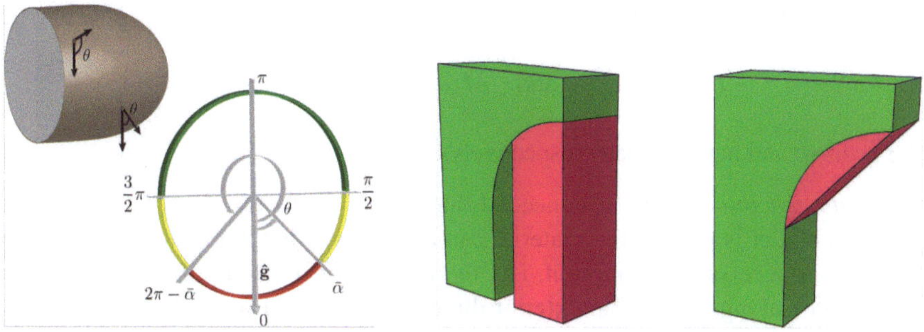

(a) (b)

Fig. 4 (**a**) Unprintable (red), modifiable (yellow) and safe (green) points with respect to the counter-clockwise angle θ between the gravity **G** and the normal **N**. Modifiable and safe points are printable. (**b**) The left grey support wastes a lot of material contrary to the chamfer on the right that saves more material and keeps the printability of the overhang as well

To this end, the surface Σ of the object Ω is divided into three subsets, on the basis of their *printability*. Denote by $\mathbf{G} = (0, 0, -1)$ the unit gravity vector, and again by $\mathbf{N}(x, y, z)$ the exterior unit normal to the surface Σ of the object Ω at the point (x, y, z). Moreover, let

$$\theta(\mathbf{N}) := \arccos\big(\mathbf{G} \cdot \mathbf{N}\big) \tag{53}$$

be the angle between **G** and **N**.

Definition 1 *A point (x, y, z) of the surface Σ is said to be*

$$\begin{aligned} &unprintable, \text{ if } \theta \in [0, \bar{\alpha}) \cup (2\pi - \bar{\alpha}, 2\pi], \\ &safe, \text{ if } \theta \in [\pi/2, 3\pi/2], \\ &modifiable, \text{ otherwise,} \end{aligned}$$

where $\bar{\alpha}$ is the given limit angle, see Fig. 4a.

While the first two definitions are immediately clear, it is worth to spend some words on the third one. Modifiable points are indeed printable since the overhang is sufficiently small. On the other hand, it could be convenient to move those points as well in order to make printable the unprintable ones. This guarantees a sufficient flexibility to shape the object conveniently and not to create long supports like the one depicted on the left of Fig. 4b. Definition 1 can be extended by saying that the set of both modifiable and safe point constitute the overall printable points.

We now focus on the construction of the vector field **v**. Recalling Eq. (48), we choose $\mathbf{v} = v\mathbf{N}$ for some scalar function v, possibly depending on **N** and κ (but not on t).

In the following we denote by

$$P(\omega) := \omega^+ \quad \text{and} \quad M(\omega) := \omega^-, \qquad \omega \in \mathbb{R},$$

the positive and negative part, respectively.

Positivity and Build Plate It is needed that $\Omega_0 \subseteq \Omega_t \subseteq \Omega_*$, for all $t \geq 0$, since once the object is printed the material can be removed but not added. This is why $v \geq 0$ is a necessary assumption, i.e., the movement of each point of the surface Σ has to be along the normal exterior direction $\mathbf{N}$. Furthermore, the object cannot move under the build plate, supposed at a fixed height $z = z_{\min} \in \mathbb{R}$. Therefore it is imposed that $v = 0$ if $z \leq z_{\min}$.

Movement of Unprintable Points We introduce the term

$$v_1(\mathbf{N}; \bar{\alpha}) := P(\cos\theta(\mathbf{N}) - \cos\bar{\alpha}), \tag{54}$$

which lets the unprintable points move outward. The speed is higher whenever θ is close to 0, which represents the (hardest) case of a horizontal hanging part.

Rotation It is convenient introducing a rotational effect in the evolution which avoids the unprintable regions to evolve "as it is" until they touch the build plate. To this end the term v_1 is multiplied by $(z_{\max} - z)$, where $z_{\max} \in \mathbb{R}$ is the maximal height reached by the object. This term simply increases the speed of lower points with respect to higher ones. This makes the lower parts be resolved (or eventually touch the built plate) before the higher parts, thus saving material.

Movement of Modifiable Points Modifiable points are moved, if necessary, by means of the following term in the vector field

$$v_2(\kappa) := M(\kappa). \tag{55}$$

It moves outward the points with negative curvature until it vanishes, i.e., the surface is locally flat. In particular, it moves concave corners and let modifiable points become a suitable support for the still unprintable points above.

Blockage of Safe Points Finally, it is necessary to exclude from the evolution the safe points of the object. In order to identify them, we use the sign of the third component n_3 of the unit exterior normal vector $\mathbf{N}$.

Final Model By putting together all the terms one obtains

$$v(x, y, z, \mathbf{N}, \kappa; \bar{\alpha}) :=$$

$$\begin{cases} C_1\,(z_{\max} - z)v_1(\mathbf{N}; \bar{\alpha}) + C_2\, v_2(\kappa), & \text{if } n_3 < 0 \text{ and } z > z_{\min}, \\ 0, & \text{otherwise,} \end{cases} \tag{56}$$

with $C_1, C_2 > 0$ positive constants (model parameters). The result expected from a such vector field is an evolution similar to the one depicted on the right in Fig. 4b, corresponding to a support whereby the angle θ in each of its point is less or equal to $\bar{\alpha}$.

The surface evolution must be stopped at some final time $t_f > 0$. It is convenient to check directly (at every time $t < t_f$) whether the overall surface is printable or not, according to Definition 1, rather than waiting that the velocity field vanishes completely. More precisely, the evolution is stopped when all the points belonging to the zero level-set are safe or modifiable, i.e., printable.

By construction, the surface always evolves towards a printable object. Indeed, any non-printable part of the surface is forced to move downward, and the surface has to stop once the build plate is reached. Nevertheless, we have no guarantee that the final object is "optimal" in terms of additional printing material. In the worst-case scenario the surface evolves until it touches the build plate, obtaining something similar to the results depicted on the left in Fig. 4b. Nevertheless, the method works fine in most cases, see [16] for some numerical results.

6.3 *Fixing Overhangs via Level-set Method 2: Topological Optimization with Shape Derivatives*

Here we present very briefly the method proposed by Allaire et al. in [8] (see also the related papers [7, 9] and [4, 5, 10].

General Ideas A different way to construct the vector field v is to adopt the strategy introduced in [6] based of the computation of the shape derivatives [3, 72]. In this approach, the user must suitably define a shape functional $J(\Omega)$ which maps any feasible subset of $\mathbb{R}^3$ to a real number. The subset Ω represents the solid surface one can possibly create, and $J(\Omega)$ somehow measures the "cost" of that object. The functional J is therefore defined in such a way as to penalize the undesired features of the surface. In the case of interest, the feature to penalize is the presence of overhangs and the computational strategy consists of finding a suitable scalar velocity field v which drives the evolution of the shape Ω_t (using the level-set method) in such a way that the function $t \mapsto J(\Omega_t)$ is decreasing. At the end of the shape optimization procedure we are hopefully left with a fully printable object Ω_*, possibly different from the original one and then, as we did in Sect. 6.2 we just need to remove the excess part $\Omega_* \backslash \Omega_0$.

Roughly speaking, the idea is explained as follows (see [3, 6] for a complete discussion and mathematical details): let $\mathbf{w}$, with $|\mathbf{w}| < 1$, be a given vector field and let O a given bounded domain of $\mathbb{R}^3$. Interpreting $\mathbf{w}$ as a displacement of the domain O, we get the new domain $O^{\mathbf{w}}$ defined by

$$O^{\mathbf{w}} := \{(x, y, z) + \mathbf{w}(x, y, z) \ : \ (x, y, z) \in O\}.$$

By definition, if the functional J is shape differentiable at O in the direction $\mathbf{w}$ the following expansion holds

$$J(O^{\mathbf{w}}) = J(O) + J'(O)(\mathbf{w}) + o(\mathbf{w}), \tag{57}$$

where the function $J'(O)(\mathbf{w})$ denotes the shape derivative of J in the direction $\mathbf{w}$.

Whenever the expression of the shape derivative falls in the general form

$$J'(O)(\mathbf{w}) = \int_{\partial O} \pi(x, y, z)\, \mathbf{w}(x, y, z) \cdot \mathbf{N}(x, y, z)\, ds \tag{58}$$

for some scalar function π, we can easily get some useful information about the corresponding shape optimization problem. Indeed, by choosing as displacement direction $\mathbf{w}$ a vector field defined as

$$\mathbf{w} = -\pi \mathbf{N},$$

we immediately get

$$J'(O)(\mathbf{w}) = -\int_{\partial O} \pi^2(x, y, z)\, ds < 0.$$

By (57), we deduce that any displacement in the direction $\mathbf{w} = -\pi\mathbf{N}$ lead to a deformation of the domain which *lowers* the value of J.

Example It is useful to note that the shape derivative can be explicitly computed in some cases. If we consider the shape functional

$$J(O) = \int_O \pi(x, y, z) dxdydz$$

for some scalar function π (in the particular case $\pi \equiv 1$, J gives the volume of the domain), the shape derivative is exactly (58); see [3, Prop. 6.22] for the detailed derivation. ■

Now, recalling how the level-set method describes the front evolution $t \to \Omega_t$ (see Eq. (48)), we get that transporting φ by the dynamics

$$\partial\varphi_t - \pi|\nabla\varphi| = 0 \tag{59}$$

is equivalent to move the boundary $\partial\Omega_t$ (i.e., the zero level-set of $\varphi(t)$) along the descent gradient direction $-J'(\Omega_t)$, thus evolving towards a lower cost shape.

Definition of J In [6] the authors propose a functional which is proven to be effective in minimizing overhangs. The idea is the following: first, let us denote by H the height of the object. Given $h \in [0, H]$, we assume that the surface was already built until height h, and now the layer at height h is being processed. Let us consider now that the only force acting on the surface Σ of the object Ω is the gravity, which pushes down the object slightly deforming its shape. The elastic displacement $\mathbf{u}_h(x, y, z) \in \mathbb{R}^3$ due to this deformation can be computed solving the well-known equation

$$\begin{cases} -\nabla \cdot \sigma(\mathbf{u}_h) = \mathbf{g}, & (x, y, z) \in \Omega \\ \mathbf{u}_h = \mathbf{0}, & (x, y, z) \in \Sigma_\downarrow \\ \sigma(\mathbf{u}_h)\mathbf{N} = \mathbf{0}, & (x, y, z) \in \Sigma \backslash \Sigma_\downarrow \end{cases} \tag{60}$$

where $\Sigma_\downarrow$ is the contact region between Ω and the build table, $\mathbf{g}(x, y, z)$ is the body force (in our case, the gravity), $\sigma(\mathbf{u}) \in \mathbb{R}^{3\times 3}$ is the matrix defined as

$$\sigma(\mathbf{u}) := \lambda(\nabla \cdot \mathbf{u})\mathbf{I} + \mu\left(\nabla\mathbf{u} + (\nabla\mathbf{u})^T\right),$$

λ and μ are the Lamé coefficients of the material, and $\mathbf{I}$ is the 3×3 unity matrix. Moreover, we have defined

$$\nabla \cdot \mathbf{u} := u_x^1 + u_y^2 + u_z^3$$

$$\nabla\mathbf{u} := \begin{pmatrix} u_x^1 & u_y^1 & u_z^1 \\ u_x^2 & u_y^2 & u_z^2 \\ u_x^3 & u_y^3 & u_z^3 \end{pmatrix}$$

$$\nabla \cdot \sigma := \begin{pmatrix} \sigma_x^{11} + \sigma_y^{12} + \sigma_z^{13} \\ \sigma_x^{21} + \sigma_y^{22} + \sigma_z^{23} \\ \sigma_x^{31} + \sigma_y^{32} + \sigma_z^{33} \end{pmatrix}.$$

Given the displacement $\mathbf{u}_h$, it is easy to compute the *compliance*, i.e., the work done by the forces acting on the object. In our case we have

$$c_h = \int_\Sigma \mathbf{g} \cdot \mathbf{u}_h ds. \tag{61}$$

Minimizing the compliance corresponds to maximize the rigidity of the object with respect to the forces acting on it. In our case this rigidity is automatically translated in a minimization of the overhangs, since deformations are greatest precisely along the protrusions.

Since overhangs should be minimized over the entire surface from bottom to top, the final cost functional J is defined by summing the compliance over all layers

$$J(\Omega) = \int_0^H c_h \, dh. \tag{62}$$

Surprising enough, the shape derivative of J can be computed explicitly and falls in the general form (58), with a suitable function π which depends on $\mathbf{u}_h$, see [8] for details.

In conclusions, one can find an overhang-free surface by evolving the initial surface Σ_0 (enclosing the domain Ω_0) in the direction $-\pi(\mathbf{u}_h)$ by means of the level-set Eq. (59), considering that at each time t the vector $\mathbf{u}_h$ must be recomputed solving (60) over the time-variable domain Ω_t (enclosed in Σ_t).

7 Building Object-Dependent Infill Structures

Besides difficulties related to 3D-printing overhanging parts, another issue arises when one has to manufacture an object: what are *inside* the surface? In principle, the ideal choice would be to print fully solid objects. In this way, the object has maximal rigidity and the internal overhanging parts are resolved (e.g., the top half of the surface of a sphere, which should lie on a support inside the sphere in order not to fall down). Unfortunately, printing fully solid objects is often not convenient because of the large quantity of material to be used. Shape optimization tools can give the optimal way to hollow out the object, reducing the overall material volume and keeping at the same time the desired rigidity and printable features. The problem reduces to finding the optimal inner structure supporting the whole object from the inside.

Commercial software typically creates infill structures with a specific pattern (e.g., squares, honeycomb, etc.) which are independent of the object under consideration. This is an easy and fast solution, but it is clearly nonoptimal is some cases, see, e.g., Fig. 5.

In the following we propose a method inspired by Dhanik and Xirouchakis [23], and Kimmel and Bruckstein [38] to create an object-dependent optimal infill

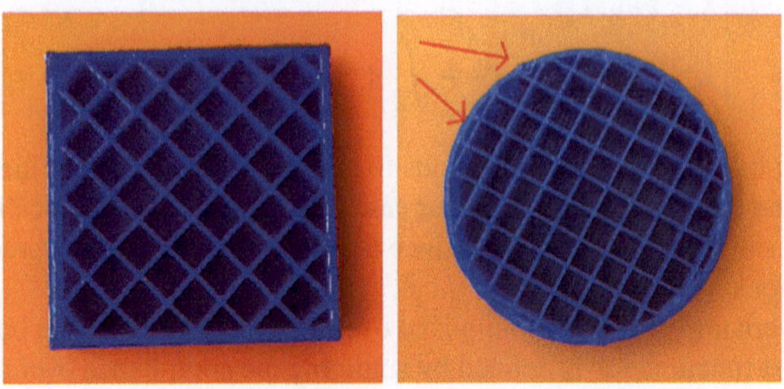

Fig. 5 Predefined square pattern for infill structure. In some parts the infill lines are very close to the boundary (external surface) of the object and/or the lines are very close to each other. As a consequence, the 3D printer hardly resolves the pattern, the extruder wastes time running for very short distance and the material already deposed melts again

(a) (b) (c)

(d) (e) (f)

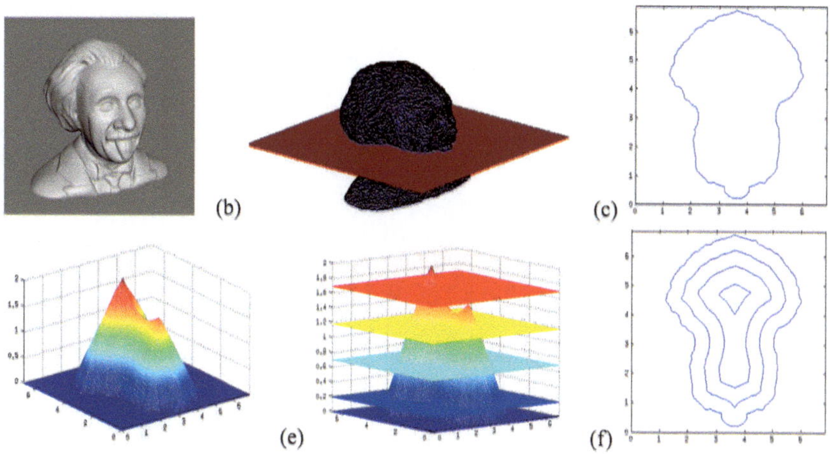

Fig. 6 (**a**) Original 3D object to be printed. (**b**) Object cut at height h (layer). (**c**) Contour Γ_h of the layer. (**d**) Distance function from the contour Σ_h, solution to the eikonal equation. (**e**) level-sets of the distance function. (**f**) level-sets as infill structure of the layer

structure specifically designed to follow the contours of the layer to be printed. This approach minimizes the changes of direction of the extruder and fix the problem shown in Fig. 5, avoiding to create holes of different size.

The method is again addressed to 3D printers with FDM technology, in which the object is created layer by layer starting from the lowest one, and it is again based on a front propagation problem and the level-set method, see Sect. 5.2. Differently from what we did before, here the level-set problem will be set in 2D (with the level-set function φ defined in $\mathbb{R}^+ \times \mathbb{R}^2$).

The idea is described in Fig. 6 and can be summarized as follows:

(a) Start from the surface to be printed, empty inside, and given in terms of facets (little triangles which cover the whole surface without leaving holes) with oriented vertices to distinguish the inner and the outer space.
(b) Cut the surface with a horizontal plane at height h. This gives the contour of the object and the layer to be printed at the current step.
(c) Denote the layer as Γ_h. This is the initial front (1D, embedded in $\mathbb{R}^2$) we want to evolve through the level-set equation in the inner normal direction $-\mathbf{N}$ until it shrinks to a single point.
(d) Solve the isotropic stationary eikonal equation $|\nabla T_h(x, y)| = 1$ inside Γ_h, with $T_h(x, y) = 0$ for $(x, y) \in \Gamma_h$. This gives the signed distance function from Γ_h [61].
(e) Cut the solution at different heights, the number of levels depends on how fine we want the infill structure.
(f) The level-sets of the solution correspond to the desired infill structure, which, by construction, follows the boundary of the object.

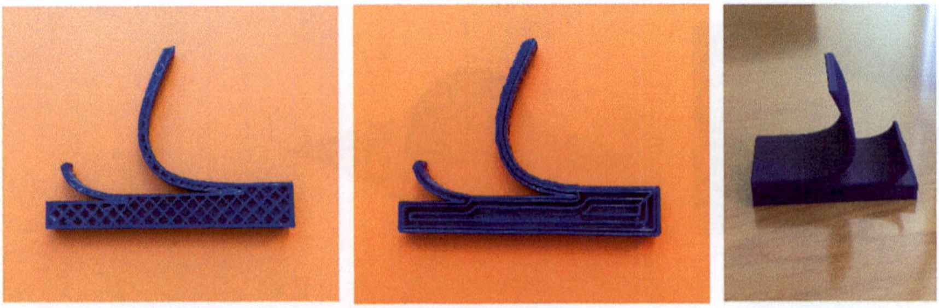

(a) (b) (c)

Fig. 7 A 3D-printed phone holder. (**a**) square infill, (**b**) proposed infill, (**c**) final object

Table 1 Comparison between the square infill and the proposed one in terms of printing time and extruded material. 3D printer was Delta WASP 2040

	SQUARE	PROPOSED	Difference
Time	18 m 30 s	11 m 24 s	-38%
Material	1559 mm	1420 mm	-9%

The procedure is then repeated for each layer until the top of the object is reached.

In Fig. 7 we show a real application of the method described above, 3D-printing a phone holder. It is basically a parallelepiped with two thin curved branches.

We have compared the eikonal-based infill with the one obtained by a predefined square pattern. The G-code (see Appendix B) was directly coded, without resorting to a commercial slicer. It is clear that with such thin branches the new method can have an advantage over a general-purpose one, derived by following the original shape of the object's boundaries. Therefore we expect a mild advantage in terms of extruded material and a more important advantage in terms of printing time. In Table 1 we summarized the results which indeed confirm the expectations.

8 Conclusions

In this chapter we have examined various mathematical models and techniques that can be applied to 3D reconstruction and 3D printing and that allow to deal with both problems in a unified manner. In particular, we have seen that nonlinear PDEs appear for the modeling of complex computer vision problems ranging from the classical case of a single image with orthographic projection under the Lambertian assumption to more general surfaces and to the multi-view problem. The starting point in the simplest case is the eikonal equation but the difficulty rapidly increases to deal with systems of nonlinear PDEs.

As for the 3D printing, we have examined some technical problems that can be solved via the level-set method for front propagation and its generalization in the context of shape optimization. Again, similar equations appear and allow to solve at least two important problems for 3D printing: the detection and resolution of overhangs, and the optimization of the deposition path to actually build the object.

We hope that this chapter will attract the attention of young researchers to this new area, which still presents many open problems as well as an increasing importance for industrial manufacturing of small and even large objects.

Acknowledgments The authors want to acknowledge the contribution of F. Tanzilli for the experiments presented in Sect. 7. The authors are members of the INdAM Research Group GNCS. This work was carried out within the INdAM-GNCS project "Metodi numerici per l'imaging: dal 2D al 3D", Code CUP_E55F22000270001.

Appendix A: The STL Format

STL (STereo Lithography interface format or Standard Triangulation Language) is the most common file format used to store object data which have to be 3D-printed. It comes in two flavors, ASCII or binary. The latter requires less space to be stored and it can be easily created from the former via free software.

Objects are determined by means of their surface, which must be closed, so to be watertight. It is also mandatory that the "interior" and the "exterior" of the solid are always defined. The surface of the solid to be printed must be tessellated by means of a triangulation. Each triangle, commonly called *facet*, is uniquely identified by the (x, y, z) coordinates of the three vertices $(\mathbf{V}^1, \mathbf{V}^2, \mathbf{V}^3)$ and by its unit exterior normal (N_x, N_y, N_z). The total number of data for each facet is 12. Moreover,

1. Each triangle must share two vertices with every neighboring triangle. In other words, a vertex of one triangle cannot lie on the side of another triangle.
2. All vertex coordinates must be strictly positive numbers. The STL file format does not contain any scale information, the coordinates are in arbitrary units. Actual units (mm, cm, ...) will be chosen in the printing process.
3. Each facet is part of the boundary between the interior and the exterior of the object. The orientation of the facets is specified redundantly in two ways. First, the direction of the normal is outward. Second, the vertices are listed in counter-clockwise order when looking at the object from the outside (right-hand rule).

The actual format of the STL file is given in the following.

```
solid name-of-the-solid
  facet normal nx ny nz
    outer loop
      vertex V1x V1y V1z
      vertex V2x V2y V2z
      vertex V3x V3y V3z
    endloop
  endfacet
```

```
  facet normal nx ny nz
    outer loop
      vertex V1x V1y V1z
      vertex V2x V2y V2z
      vertex V3x V3y V3z
    endloop
  endfacet
  [...]
endsolid name-of-the-solid
```

Values are *float* numbers and indentation is made by two blanks (no tab). Unit normal direction can be simply computed by

$$\mathbf{N} = \frac{(\mathbf{V}^2 - \mathbf{V}^1) \times (\mathbf{V}^3 - \mathbf{V}^2)}{|(\mathbf{V}^2 - \mathbf{V}^1) \times (\mathbf{V}^3 - \mathbf{V}^2)|}.$$

Appendix B: The G-code Format

The STL file described above is far from being processed by a 3D printer. Another step is needed in order to transform the set of facets composing the surface of the object in something which can be understood by the machine.

The G-code is a programming language for CNC (Computer Numerical Control) machines and 3D printers. In the case of 3D-printers, it is used to tell the machine the exact path the extruder should follow, how fast to move and how much material should be deposed. Putting the infill problem aside for a moment, the extruder's path is obtained by slicing the triangulated surface by horizontal planes and taking the intersection between the facets and each plane, see Fig. 6a–c. The resulting curve is then approximated by a sequence of points. The machine will pass through these points in the given order, extruding the material.

The core of a G-code file sounds like this:

```
G1 F# X# Y# Z# E#
G1 F# X# Y# Z# E#
G1 F# X# Y# Z# E#
[...]
```

where ♯ is a *float* number. G1 means that the extruder has to move along a straight line (circular patterns are also possible with G2 and G3), F is the speed of the extruder, the following triple X, Y, Z represents the coordinates of the point to be reached, in this step, from the current position of the extruder. Finally, E indicates how much material has to be extruded up to that line of code.

Many other commands exist and are used to start/stop the machine, set the units of measure, set the temperature of the nozzle/bed, set fan speed, etc.

References

1. Ahmed, A.H., Farag, A.A.: A new formulation for shape from shading for Non-Lambertian surfaces. In: Proceeding of the IEEE Conference on Computer Vision and Pattern Recognition (CVPR), pp. 1817–1824 (2006)
2. Ahmed, A.H., Farag, A.A.: Shape from shading under various imaging conditions. In: IEEE International Conference on Computer Vision and Pattern Recognition CVPR'07, pp. X1–X8. IEEE, Minneapolis, MN (2007)
3. Allaire, G.: Conception Optimale de Structures. Springer, Berlin Heidelberg (2007)
4. Allaire, G., Bogosel, B.: Optimizing supports for additive manufacturing. Struct. Multidiscip. Optim. **58**, 2493–2515 (2018)
5. Allaire, G., Jakabčin, L.: Taking into account thermal residual stresses in topology optimization of structures built by additive manufacturing. Math. Models Methods Appl. Sci. **28**, 2313–2366 (2018)
6. Allaire, G., Jouve, F., Toader, A.-M.: Structural optimization using sensitivity analysis and a level-set method. J. Comput. Phys. **194**, 363–393 (2004)
7. Allaire, G., Dapogny, C., Estevez, R., Faure, A., Michailidis, G.: Structural optimization under overhang constraints imposed by additive manufacturing processes: an overview of some recent results. Appl. Math. Nonlinear Sci. **2**(2), 385–402 (2017)
8. Allaire, G., Dapogny, C., Estevez, R., Faure, A., Michailidis, G.: Structural optimization under overhang constraints imposed by additive manufacturing technologies. J. Comput. Phys. **351**, 295–328 (2017)
9. Allaire, G., Dapogny, C., Faure, A., Michailidis, G.: Shape optimization of a layer by layer mechanical constraint for additive manufacturing. C. R. Acad. Sci. Paris, Ser. I **355**(6), 699–717 (2017)
10. Allaire, G., Bihr, M., Bogosel, B.: Support optimization in additive manufacturing for geometric and thermo-mechanical constraints. Struct. Multidiscip. Optim. **61**, 2377–2399 (2020)
11. Attene, M.: Shapes in a box: disassembling 3D objects for efficient packing and fabrication. Comput. Graphics Forum **34**, 64–76 (2015)
12. Barles, G.: Solutions de viscosité des équations de Hamilton-Jacobi. In: Mathématiques et Applications, vol. 17. Springer, Berlin (1994)
13. Bartoli, A., Gérard, Y., Chadebecq, F., Collins, T., Pizarro, D.: Shape-from-Template. IEEE Trans. Pattern Anal. Mach. Intell. **37**(10), 2099–2118 (2015)
14. Blinn, J.F.: Models of light reflection for computer synthesized pictures. Comput. Graph. **11**(2), 192–198 (1977)
15. Breuß, M., Cristiani, E., Durou, J.-D., Falcone, M., Vogel, O.: Perspective shape from shading: ambiguity analysis and numerical approximations. SIAM J. Imag. Sci. **5**(1), 311–342 (2012)
16. Cacace, S., Cristiani, E., Rocchi, L.: A level set based method for fixing overhangs in 3D printing. Appl. Math. Model. **44**, 446–455 (2017)
17. Camilli, F., Siconolfi, A.: Maximal subsolutions for a class of degenerate Hamilton-Jacobi problems. Indiana Univ. Math. J. **48**(3), 1111–1131 (1999)
18. Camilli, F., Tozza, S.: A unified approach to the well-posedness of some non-Lambertian models in Shape-from-Shading theory. SIAM J. Imag. Sci. **10**(1), 26–46 (2017)
19. Chambolle, A.: A uniqueness result in the theory of stereo vision: coupling Shape from Shading and binocular information allows unambiguous depth reconstruction. Annales de l'Istitute Henri Poincaré **11**(1), 1–16 (1994)
20. Chung, H.-S., Jia, J.: Efficient photometric stereo on glossy surfaces with wide specular lobes. In: IEEE Conference on Computer Vision and Pattern Recognition, 2008 (CVPR 2008), pp. 1–8 (2008)
21. Cook, R.L., Torrance, K.E.: A reflectance model for computer graphics. ACM Trans. Graph. **1**(1), 7–24 (1982)

22. Crandall, M.G., Ishii, H., Lions, P.-L.: Uniqueness of viscosity solutions of Hamilton-Jacobi equations revisited. J. Math. Soc. Jpn. **39**(4), 581–596 (1987)
23. Dhanik, S., Xirouchakis, P.: Contour parallel milling tool path generation for arbitrary pocket shape using a fast marching method. Int. J. Adv. Manuf. Technol. **50**, 1101–1111 (2010)
24. Durou, J.-D., Courteille, F.: Integration of a normal field without boundary condition. In: Proceedings of the First International Workshop on Photometric Analysis For Computer Vision-PACV 2007, pp. 8–p. INRIA, France (2007)
25. Durou, J.-D., Aujol, J.-F., Courteille, F.: Integrating the normal field of a surface in the presence of discontinuities. In: Energy Minimization Methods in Computer Vision and Pattern Recognition (EMMCVPR), vol. 5681, pp. 261–273 (2009)
26. Durou, J.-D., Falcone, M., Sagona, M.: Numerical methods for shape-from-shading: A new survey with benchmarks. Comput. Vis. Image Underst. **109**(1), 22–43 (2008)
27. Durou, J.-D., Falcone, M., Quéau, Y., Tozza, S.: A Comprehensive Introduction to Photometric 3D-Reconstruction, pp. 1–29. Springer International Publishing, Cham (2020)
28. El-Sayegh, S., Romdhane, L., Manjikian, S.: A critical review of 3D printing in construction: benefits, challenges, and risks. Archiv. Civ. Mech. Eng. **20**, 34 (2020)
29. Ezair, B., Massarwi, F., Elber, G.: Orientation analysis of 3D objects toward minimal support volume in 3D-printing. Comput. Graph. **51**, 117–124 (2015)
30. Falcone, M., Ferretti, R.: Semi-Lagrangian approximation schemes for linear and Hamilton-Jacobi equations. In: Society for Industrial and Applied Mathematics, 1st edn. (2014)
31. Favaro, P., Soatto, S.: A geometric approach to shape from defocus. IEEE Trans. Pattern Anal. Mach. Intell. **27**(3), 406–417 (2005)
32. Georghiades, A.S.: Incorporating the Torrance and Sparrow model of reflectance in uncalibrated photometric stereo. In: Proceedings of the Ninth IEEE International Conference on Computer Vision (ICCV 2003), vol. 2, pp. 816–823. IEEE, New York (2003)
33. Healey, G., Binford, T.O.: Local shape from specularity. In: Department of Computer Science, Stanford University, Artificial Intelligence Laboratory (1986)
34. Horn, B.K.P.: Shape from Shading: A Method for Obtaining the Shape of a Smooth Opaque Object from One View. PhD Thesis, MIT, New York (1970)
35. Horn, B.K.P., Brooks, M.J.: The variational approach to shape from shading. Computer Vision, Graphics, and Image Processing **33**(2), 174–208 (1986)
36. Jiang, J., Ma, Y.: Path planning strategies to optimize accuracy, quality, build time and material use in additive manufacturing: A review. Micromachines **11**, 633 (2020)
37. Jin, Y.-A., He, Y., Fu, J.-Z., Gan, W.-F., Lin, Z.-W.: Optimization of tool-path generation for material extrusion-based additive manufacturing technology. Addit. Manuf. **1–4**, 32–47 (2014)
38. Kimmel, R., Bruckstein, A.M.: Shape offsets via level sets. Comput. Aided Des. **25**, 154–162 (1993)
39. Kruzkov, S.N.: The generalized solution of the Hamilton-Jacobi equations of eikonal type I. Math. USSR Sbornik **27**, 406–446 (1975)
40. Lan, Y., Dong, Y., Pellacini, F., Tong, X.: Bi-scale appearance fabrication. ACM Trans. Graph. **32**(4), 145:1–145:12 (2013)
41. Langelaar, M.: Combined optimization of part topology, support structure layout and build orientation for additive manufacturing. Struct. Multidiscip. Optim. **57**, 1985–2004 (2018)
42. Lions, P.-L.: Generalized Solutions of Hamilton-Jacobi Equations. Pitman, London (1982)
43. Lions, P.-L., Rouy, E., Tourin, A.: Shape-from-shading, viscosity solutions and edges. Numer. Math. **64**(1), 323–353 (1993)
44. Livesu, M., Ellero, S., Martínez, J., Lefebvre, S., Attene, M.: From 3D models to 3D prints: an overview of the processing pipeline. Comput. Graph. Forum **36**, 537–564 (2017)
45. Mecca, R., Falcone, M.: Uniqueness and approximation of a photometric shape-from-shading model. SIAM J. Imag. Sci. **6**(1), 616–659 (2013)
46. Mecca, R., Tozza, S.: Shape Reconstruction of Symmetric Surfaces Using Photometric Stereo, pp. 219–243. Springer, Berlin (2013)

47. Ngan, A., Durand, F., Matusik, W.: Experimental analysis of BRDF models. In: Bala, K., Dutre, P. (eds.) Eurographics Symposium on Rendering (2005). The Eurographics Association, New York (2005)
48. Onn, R., Bruckstein, A.M.: Integrability disambiguates surface recovery in two-image photometric stereo. Int. J. Comput. Vis. **5**(1), 105–113 (1990)
49. Oren, M., Nayar, S.K.: Generalization of Lambert's reflectance model. In: SIGGRAPH '94 Proceedings of the 21st Annual Conference on Computer Graphics and Interactive Techniques, Orlando, FL, USA, pp. 239–246 (1994)
50. Oren, M., Nayar, S.K.: Generalization of the Lambertian model and implications for machine vision. Int. J. Comput. Vis. **14**(3), 227–251 (1995)
51. Osher, S., Fedkiw, R.: Level set methods and dynamic implicit surfaces. In: Applied Mathematical Sciences, vol. 153. Springer, New York (2003)
52. Osher, S., Sethian, J.A.: Front propagating with curvature-dependent speed: algorithms based on Hamilton-Jacobi formulations. J. Comput. Phys. **79**(1), 12–49 (1988)
53. Phong, B.T.: Illumination for computer generated pictures. Commun. ACM **18**(6), 311–317 (1975)
54. Piovarči, M., Wessely, M., Jagielski, M.J., Alexa, M., Matusik, W., Didyk, P.: Directional screens. In: Proceedings of the 1st Annual ACM Symposium on Computational Fabrication (2017)
55. Prados, E., Faugeras, O.: A mathematical and algorithmic study of the Lambertian SFS problem for orthographic and pinhole cameras. In: Rapport de recherche 5005, Institut National de Recherche en Informatique et en Automatique, Sophia Antipolis, France (2003)
56. Prados, E., Faugeras, O.: Shape from shading: a well-posed problem? In: International Conference on Computer Vision and Pattern Recognition CVPR05, vol. 2, pp. 870–877. IEEE, San Diego (2005)
57. Prados, E., Faugeras, O., Camilli, F.: Shape from shading: a well-posed problem? Technical Report 5297, INRIA, New York (2004)
58. Quéau, Y., Durou, J.-D., Aujol, J.-F.: Normal integration: a survey. J. Math. Imaging Vision **60**(4), 576–593 (2018)
59. Quéau, Y., Durou, J.-D., Aujol, J.-F.: Variational methods for normal integration. J. Math. Imaging Vis. **60**(4), 609–632 (2018)
60. Rouiller, O., Bickel, B., Kautz, J., Matusik, W., Alexa, M.: 3D-printing spatially varying BRDFs. IEEE Comput. Graph. Appl. **33**(6), 48–57 (2013)
61. Sethian, J.A.: Level Set Methods and Fast Marching Methods: Evolving Interfaces in Computational Geometry, Fluid Mechanics, Computer Vision, and Material Science. Cambridge University Press, New York (1999)
62. Tay, Y.W.D., Panda, B., Paul, S.C., Noor Mohamed, N.A., Tan, M.J., Leong, K.F.: 3D printing trends in building and construction industry: a review. Virtual and Physical Prototyping **12**, 261–276 (2017)
63. Torrance, K.E., Sparrow, E.M.: Theory for off-specular reflection from roughened surfaces. Josa **57**(9), 1105–1114 (1967)
64. Tozza, S.: Perspective shape-from-shading problem: a unified convergence result for several non-Lambertian models. J. Imaging **8**(2), 36 (2022)
65. Tozza, S., Falcone, M.: Analysis and approximation of some shape-from-shading models for non-Lambertian surfaces. J. Math. Imaging Vision **55**(2), 153–178 (2016)
66. Tozza, S., Falcone, M.: A comparison of non-Lambertian models for the shape-from-shading problem. In: Breuß, M., Bruckstein, A., Maragos, P., Wuhrer, S. (eds.) Perspectives in Shape Analysis, pp. 15–42. Springer International Publishing, Cham (2016)
67. Tozza, S., Mecca, R., Duocastella, M., Del Bue, A.: Direct differential photometric stereo shape recovery of diffuse and specular surfaces. J. Math. Imaging Vision **56**(1), 57–76 (2016)
68. van de Ven, E., Ayas, C., Langelaar, M., Maas, R., van Keulen, F.: A PDE-based approach to constrain the minimum overhang angle in topology optimization for additive manufacturing.

In: Schumacher, A., Vietor, T., Fiebig, S., Bletzinger, K.-U., Maute, K. (eds.) Advances in Structural and Multidisciplinary Optimization (WCSMO 2017), pp. 1185–1199. Springer, Cham (2018)

69. van de Ven, E., Maas, R., Ayas, C., Langelaar, M., van Keulen, F.: Continuous front propagation-based overhang control for topology optimization with additive manufacturing. Struct. Multidiscip. Optim. **57**, 2075–2091 (2018)
70. van de Ven, E., Maas, R., Ayas, C., Langelaar, M., van Keulen, F.: Overhang control based on front propagation in 3D topology optimization for additive manufacturing. Comput. Methods Appl. Mech. Eng. **369**, 113169 (2020)
71. van de Ven, E., Maas, R., Ayas, C., Langelaar, M., van Keulen, F.: Overhang control in topology optimization: a comparison of continuous front propagation-based and discrete layer-by-layer overhang control. Struct. Multidiscip. Optim. **64**, 761–778 (2021)
72. van Dijk, N.P., Maute, K., Langelaar, M., van Keulen, F.: Level-set methods for structural topology optimization: a review. Struct. Multidiscip. Optim. **48**, 437–472 (2013)
73. van Oosterom, A., Strackee, J.: The solid angle of a plane triangle. IEEE Trans. Biomed. Eng. **BME-30**(2), 125–126 (1983)
74. Ward, G.J.: Measuring and modeling anisotropic reflection. In: Proceedings of the 19th Annual Conference on Computer Graphics and Interactive Techniques (SIGGRAPH '92), pp. 265–272. Association for Computing Machinery, New York (1992)
75. Weyrich, T., Peers, P., Matusik, W., Rusinkiewicz, S.: Fabricating microgeometry for custom surface reflectance. ACM Trans. Graph. **28**, 32, 1–6 (2009)
76. Witkin, A.P.: Recovering surface shape and orientation from texture. Artif. Intell. **17**(1–3), 17–45 (1981)
77. Wolff, L.B.: Diffuse-reflectance model for smooth dielectric surfaces. J. Opt. Soc. Am. A **11**(11), 2956–2968 (1994)
78. Wolff, L.B., Boult, T.E.: Constraining object features using a polarization reflectance model. IEEE Trans. Pattern Anal. Mach. Intell. **13**(7), 635–657 (1991)
79. Wolff, L.B., Nayar, S.K., Oren, M.: Improved diffuse reflection models for computer vision. Int. J. Comput. Vis. **30**, 55–71 (1998)
80. Yao, M., Chen, Z., Luo, L., Wang, R., Wang, H.: Level-set-based partitioning and packing optimization of a printable model. ACM Trans. Graph. **34**(6), 214:1–214:11 (2015)
81. Yuille, A.L., Snow, D., Epstein, R., Belhumeur, P.N.: Determining generative models of objects under varying illumination: shape and albedo from multiple images using SVD and integrability. Int. J. Comput. Vis. **35**(3), 203–222 (1999)
82. Zhang, R., Tsai, P.-S., Cryer, J.E., Shah, M.: Shape-from-shading: a survey. IEEE Trans. Pattern Anal. Mach. Intell. **21**(8), 690–706 (1999)
83. Zheng, Z., Ma, L., Li, Z., Chen, Z.: An extended photometric stereo algorithm for recovering specular object shape and its reflectance properties. Comput. Sci. Inf. Syst. **7**(1), 1–12 (2010)

Photometric Stereo with Non-Lambertian Preprocessing and Hayakawa Lighting Estimation for Highly Detailed Shape Reconstruction

Georg Radow, Giuseppe Rodriguez, Ashkan Mansouri Yarahmadi, and Michael Breuß

Abstract In many realistic scenarios, the use of highly detailed photometric 3D reconstruction techniques is hindered by several challenges in given imagery. Especially, the light sources are often unknown and need to be estimated, and the light reflectance is often non-Lambertian. In addition, when approaching the problem to apply photometric techniques at real-world imagery, several parameters appear that need to be fixed in order to obtain high-quality reconstructions. In this chapter, we attempt to tackle these issues by combining photometric stereo with non-Lambertian preprocessing and Hayakawa lighting estimation. At hand of a dedicated study, we discuss the applicability of these techniques for their use in automated 3D geometry recovery for 3D printing.

Keywords Photometric stereo · Shape from shading · Hayakawa procedure · Lighting estimation · Oren-Nayar model · Lambertian reflector

1 Introduction

Photometric stereo (PS) is a fundamental inverse problem aiming at reconstructing the shape of a three-dimensional object based on a set of images acquired under a varying source of light. Under this assumption, the images embed the shape and color information of the observed object. Despite its long history in computer vision [49, 50], PS is still a fundamentally challenging research problem due to the unknown reflectance and global lighting effects of real-world objects [41]. For

G. Radow · A. M. Yarahmadi · M. Breuß
BTU Cottbus-Senftenberg, Applied Mathematics Group, Institute for Mathematics, Cottbus, Germany
e-mail: Radow@b-tu.de; Yarahmadi@b-tu.de; Breuss@b-tu.de

G. Rodriguez (✉)
University of Cagliari, Department of Mathematics and Computer Science, Cagliari, Italy
e-mail: rodriguez@unica.it

E. Cristiani et al. (eds.), *Mathematical Methods for Objects Reconstruction*,
Springer INdAM Series 54, https://doi.org/10.1007/978-981-99-0776-2_2

the sake of simplicity, the lighting positions and directions are often assumed to be known across the PS research community [5, 28, 38, 48]. However, the real-world applications of PS mostly deal with data represented by images acquired under unknown light conditions; see [6, 11], and [21]. In [21], it was shown that at least 6 differently illuminated images are needed to resolve the light positioning, leading to a successful PS scenario that will be further discussed in the current work.

In an almost parallel paradigm, shape from shading (SfS) aims at solving the modeling inverse problem with a set of assumptions similar to PS, except for the fact that in SfS only a single two-dimensional image of the object is at hand [16, 52]. The common assumptions among classic (orthographic) PS and SfS are: *(i)* the illumination in the photographed scene, *(ii)* the light reflectance surface properties, and *(iii)* the orthographic projection that is performed by the camera needs to be known.

Let us first comment on the camera model employed in our setting. In this context, we note that both SfS and PS allow to employ different camera models. An important example is the use of a perspective camera projection that yields more complex equations than in the orthographic setting; see [9] for a comprehensive introduction to perspective SfS and [28, 33, 44] for examples of works on perspective PS. In this chapter, we adopt a canonical orthographic camera as it is suitable in applications with an object located at a relatively far distance from the camera, compared to the object dimensions. This holds true, in particular, in the dataset we focus on, representing a highly detailed seashell illuminated by sunlight. We have validated by undocumented tests that indeed our application scenario is represented well by the orthographic setting. As another classical choice in the setting employed in this chapter, we consider a classical illumination model [24, 25] that idealizes the environmental light as parallel beams emanating from a light source located at infinity, e.g., the Sun.

As indicated, we will follow the Hayakawa lighting estimation method [21] to compute the lighting directions and to make use of them in PS. As a related but different topic of research, the uncalibrated PS approach [6] aims to resolve lighting and 3D depth information simultaneously, which naturally leads to highly sophisticated problem formulations; see, e.g., [19] where non-convex optimization problems need to be resolved. In the context of this line of research, we also mention deep learning approaches such as the Lighting Calibration Network and its extensions [13, 14]. However, as it often happens with deep learning approaches, the mechanisms behind the estimation are still largely unclear, even if attempts for an analysis give some useful insights [14]. In this chapter, we prefer to exploit lighting information directly and to keep in this way the explainability of results. Furthermore, there is certainly a point for estimating lighting directly, if it is feasible, and to keep the whole setup more simple and tractable compared to uncalibrated PS.

We now make some remarks on the light reflected by the photographed objects. In classic works such as [24], Lambertian light reflectance is the most common model in photometric methods. Lambert's model [31] itself is an idealized model and corresponds to a very matte surface without specular highlights, limiting its application in real-world scenarios. Hence, extending PS to work with non-

Lambertian surfaces has been of interest for its practical use. Concerning the seashell dataset that we consider as a best practice example, we remark that the surface of the seashell shows a degree of roughness that can be well understood by a non-Lambertian representation [34, 35, 45], namely the Oren–Nayar, Blinn–Phong, or Torrance–Sparrow model. Among them, the Oren–Nayar model [34] appears to be a pragmatic choice, as it is a reasonable model for matte non-Lambertian surfaces that has been utilized in computer vision, cf. [27, 46, 47].

With recent advances in deep learning schemes, in contrast to more sophisticated mathematical models [34, 35, 45], a range of data-driven approaches have been introduced to model object reflectance properties [12, 26, 40]. Here, we briefly discuss the applicability of some state-of-the-art reflectance modeling approaches to our setup. The deep PS network [40] may not be applicable to our case, because of its predefined light directions assumption between the training and testing phases, while we consider the Sun as a source of light, whose changing position is unknown and needs to be estimated. The convolutional-neural-network-based PS introduced in [26] relaxes the lighting constraint in [40], but it is primarily concerned with modeling isotropic materials, namely glass and plastic, whose reflections are invariant under rotation about the surface normal. This clearly does not conform to our rough object of interest.

Along with [26], the approach [43] proposed a physics-based unsupervised learning framework to inversely render the general reflectance properties by removing the fixed light constraint. Here, a permutation of the light directions during the training and testing phases is allowed as it impacts the final reconstructed model. In general, all the methods [26, 40, 43] consider a fixed number of input channels to their learning schemes, which is another major constraint in PS.

In a general PS scenario, one may consider the varying number of input images as different channels of a spectral image, as in case of the three channels corresponding to an RGB image. This reformulation in the structure of the input images requires a flexible deep learning scheme that accepts a varying number of input images, in contrast to those schemes with a fixed number of input channels. The channel constraint is of specific importance in case a convolutional layer is chosen as the input layer of the deep learning scheme.

The model based on a deep fully convolutional network [12] resolves the constraint on the number of input channels by locating a recurrent neural network [51] as the input layer, though the lighting knowledge about the training data is still assumed to be known a priori.

One of the challenges that we face in our current study is to derive the positions and directions of the light source, represented by the Sun in an outdoor environment, based only on a set of images of the object. In [23], PS was investigated during a single day under a variety of sunlight conditions obtained from a sky probes dataset [22], but to the best of our knowledge no deep learning scheme addresses an outdoor PS scenario with unknown light sources.

In the following, we will discuss Hayakawa's method for estimating light positions [21], as well as a simplified Oren–Nayar approach [34, 39]. Their combination is the main line of the current study.

2 Mathematical Setup

A classical orthographic SfS model [29] with a varying source of light illuminating an approximately Lambertian [34] object is the main element of our setup. Let us describe them in detail.

We assume a right-handed reference system in $\mathbb{R}^3$ along with a camera placed at infinity, performing an orthographic projection such that the z-axis and the optical axis of the camera coincide. A varying light source represented by the vectors

$$\boldsymbol{\ell}_t = (\ell_{1t}, \ell_{2t}, \ell_{3t})^\top, \quad t = 1, \ldots, q,$$

illuminates the object from an infinite distance and from q different directions. Note that $\|\boldsymbol{\ell}_t\|$ is proportional to the light intensity, leading to the introduction of an undetermined proportionality constant to the problem. Here and in the following, $\|\cdot\|$ represents the *Euclidean* norm.

We assume each vector $\boldsymbol{\ell}_t$ emanates from the object to the light source and that both the normal vectors to the object surface and the light vectors themselves point to the same half-space, determined by the positive direction of the z-axis. Since we will employ the Oren–Nayar model [34] to approximate the observed object by a Lambertian one, the incident angle between the light and the normal vectors will be of our specific interest, as it will be further explained in Sect. 5.

In this setting, we capture q images, each with horizontal and vertical sides of sizes A and B, respectively. Each image is considered as the sampling of a function $u(x, y)$, defined on the domain $\Omega = [-A/2, A/2] \times [-B/2, B/2]$, at the points

$$\left(x_i, y_j\right) := (-A/2 + ih, -B/2 + jh),$$

letting $i \in \{0, \ldots, r+1\}$, $j \in \{0, \ldots, s+1\}$, $s, r \in \mathbb{N}$, $h = A/(r+1)$, and $B = (s+1)\,h$. The size of the corresponding discretized images is $(r+2) \times (s+2)$.

To each point $(x, y) \in \Omega$, a depth value $u(x, y) \in \mathbb{R}^+$ is associated, with a gradient vector

$$\nabla u\,(x, y) = \left(\frac{\partial u\,(x, y)}{\partial x}, \frac{\partial u\,(x, y)}{\partial y}\right)^\top = \left(u_x, u_y\right)^\top, \tag{1}$$

and a normal vector

$$\mathbf{n}(x, y) = \frac{\left(-u_x, -u_y, 1\right)^\top}{\sqrt{1 + \|\nabla u\|^2}}. \tag{2}$$

One of the main objects of interest in this study is the approximation to a Lambertian surface based on the Oren–Nayar model [34]. This allows one to assume that Lambert's cosine law

$$\rho(x, y)\,\langle \mathbf{n}(x, y), \boldsymbol{\ell}_t \rangle = \mathcal{I}_t(x, y), \quad t = 1, \ldots, q, \tag{3}$$

holds true, where $\langle\cdot,\cdot\rangle$ is the usual inner product in $\mathbb{R}^3$, the light intensity at each point of the tth image is denoted by $\mathcal{I}_t(x, y)$, and the scalar function $\rho(x, y)$ represents the *albedo* at each surface point and keeps into account the partial light absorption of that portion of the surface.

For what follows, it is convenient that the captured images are stored in vector form, so we order their pixels lexicographically. Here, the coordinate pixel $(x_i, y_j) \in \Omega$ is mapped to the index $k = (i-1)s + j$, where $k \in \{1, \dots, p\}$ and $p = (r+2)(s+2)$ is the number of image pixels. In the following, we will assume $p \gg q$, since the number of pixels in an image is usually very large, while we aim at obtaining a reconstruction using a small set of images. The corresponding vector images are denoted by $\mathbf{m}_1, \mathbf{m}_2, \dots, \mathbf{m}_q \in \mathbb{R}^p$, and we rewrite the discretization in any of the following forms:

$$\begin{aligned}
u(x_i, y_j) &= u_{i,j} = u_k, \\
u_x(x_i, y_j) &= (u_x)_{i,j} = (u_x)_k, \\
u_y(x_i, y_j) &= (u_y)_{i,j} = (u_y)_k, \\
\mathbf{n}(x_i, y_j) &= \mathbf{n}_{i,j} = \mathbf{n}_k, \\
\mathcal{I}_t(x_i, y_j) &= (\mathbf{m}_t)_k = m_{k,t},
\end{aligned}$$

depending on the context.

Here is a brief review of our classical assumptions:

- The surface is approximated by a Lambertian one, based on [34, 39].
- The sources of light are placed at infinite distance from the object.
- No shadow is cast on the surface.
- The camera is sufficiently far from the object, resulting in the absence of any perspective deformation.

Under these assumptions, rewriting Lambert's law (3), we obtain

$$\rho(x, y)\left\langle \frac{(-u_x, -u_y, 1)^\top}{\sqrt{1 + \|\nabla u(x, y)\|^2}}, (\ell_{1t}, \ell_{2t}, \ell_{3t})^\top \right\rangle = \mathcal{I}_t(x, y),$$

from which, multiplying both sides by $\sqrt{1 + \|\nabla u(x, y)\|^2} \geq 1$,

$$\mathcal{I}(x, y)\sqrt{1 + \|\nabla u(x, y)\|^2} - \rho(x, y)\left\langle(-u_x, -u_y, 1)^\top, (\ell_{1t}, \ell_{2t}, \ell_{3t})^\top\right\rangle = 0.$$

Next,

$$\begin{aligned}
0 &= \mathcal{I}(x, y)\sqrt{1 + \|\nabla u(x, y)\|^2} + \rho(x, y)\left(\left\langle\nabla u(x, y), (\ell_{1t}, \ell_{2t})^\top\right\rangle - \ell_{3t}\right) \\
&:= H_t(x, y, \nabla u(x, y)) \qquad (4)
\end{aligned}$$

is deduced after performing the scalar product and considering (1).

At any surface point $(x, y, u(x, y)) \in \mathbb{R}^3$, a normal vector $(-\nabla u(x, y), 1)^\top$ is approximated using (4), provided Dirichlet boundary conditions

$$u(x, y) = g(x, y), \quad \forall (x, y) \in \partial\Omega,$$

are addressed, where $\partial\Omega$ denotes the boundary of the domain Ω. In conclusion, one obtains the following Hamilton–Jacobi differential model:

$$\begin{cases} H_t(x, y, \nabla u(x, y)) = 0, & t = 1, \ldots, q, \\ u(x, y) = g(x, y), & (x, y) \in \partial\Omega. \end{cases} \tag{5}$$

3 Photometric Stereo with Known Lighting

We recall that, by enforcing the Dirichlet boundary condition on (5), we aim at solving a nonlinear system of q first-order partial differential equations of Hamilton–Jacobi type. Assuming that the light directions $\boldsymbol{\ell}_t, t = 1, \ldots, q$, are known and following [32], we let $t = 1$ in (3) to obtain

$$\sqrt{1 + \|\nabla u(x, y)\|^2} = \rho(x, y) \frac{\langle -\nabla u(x, y), \tilde{\boldsymbol{\ell}}_1 \rangle + \ell_{31}}{\mathcal{I}_1(x, y)}, \tag{6}$$

with $\tilde{\boldsymbol{\ell}}_t := (\ell_{1t}, \ell_{2t})^\top \in \mathbb{R}^2$. Next, we substitute (6) in the equations corresponding to $t = 2, \ldots, q$, obtaining

$$\left(\langle -\nabla u(x, y), \tilde{\boldsymbol{\ell}}_1 \rangle - \ell_{31}\right) \mathcal{I}_t(x, y) = \left(\langle -\nabla u(x, y), \tilde{\boldsymbol{\ell}}_t \rangle - \ell_{3t}\right) \mathcal{I}_1(x, y). \tag{7}$$

The well-posedness of (7) is assured in the case $q \geq 2$, that is, if at least two images are available; see also [36]. In practice, the bigger is q, the more likely we are able to construct a meaningful solution. Indeed, when $q = 2$, the solution may not be inferred if the assumptions on the method are not perfectly verified. On the contrary, a dataset with three or more images illuminated from accurately known light sources leads to a least-square problem that effectively reduces the noise influence and results in a better approximation of the depth function $u(x, y)$.

Finally, the albedo is found by the equation

$$\rho(x, y) = \frac{\mathcal{I}_t(x, y)}{\langle \mathbf{n}(x, y), \boldsymbol{\ell}_t \rangle}, \quad \text{for any } t = 1, \ldots, q.$$

The conditions for the existence and uniqueness of the solution of (5) have been studied in [30]. The reader is further referred to [32] for a study of the problem at hand with a set of more realistic assumptions. In addition, [42] provides a fundamental treatment of (5) in the PS context.

We stress once again the importance of the lights position, since in certain illuminating conditions the coefficient matrix of the linear system resulting from the discretization of (5) might be singular or severely ill-conditioned. However, a suitable positioning of light sources can resolve the issue.

The above approach is a global one. Alternatively, the classical approach [50] for solving PS is to replace the product of albedo and normal vector with a single vector $\tilde{\mathbf{n}} := \rho\mathbf{n}$. In the following, we confine our attention to the subset $\tilde{\Omega} \subset \Omega$, which corresponds to those locations belonging to the actual object that we want to reconstruct, and not to the background. Then, the solution for each $(x, y) \in \tilde{\Omega}$ is found locally via

$$\hat{\mathbf{n}} = \arg\min_{\tilde{\mathbf{n}} \in \mathbb{R}^3} \sum_{t=1}^{q} \left(\langle \tilde{\mathbf{n}}, \boldsymbol{\ell}_t \rangle - \mathcal{I}_t(x, y) \right)^2, \quad \rho(x, y) = \left\| \hat{\mathbf{n}} \right\|, \quad \mathbf{n}(x, y) = \frac{\hat{\mathbf{n}}}{\left\| \hat{\mathbf{n}} \right\|}. \tag{8}$$

Contrary to the previous approach, the minimization in (8) yields a unique solution if $q \geq 3$ and three light sources must be non-coplanar, cf. [50].

The normal vectors obtained through (8) are the best local explanation of the sampled images with known lighting, according to Lambert's reflectance model and without considering robustness to noise. However, in general, they are not integrable, i.e., they do not correspond to an actual surface. Then, the second step consists of obtaining the depth u through numerical normal integration. Following [37], one approach is to compute u as

$$u = \arg\min_{u} \iint_{\tilde{\Omega}} \|\nabla u(x, y) - g_{\perp}(x, y)\|^2 \, \mathrm{d}x\mathrm{d}y \quad \Longleftrightarrow \quad \Delta u = \mathrm{div} g_{\perp}. \tag{9}$$

We first discuss the construction of $g_{\perp} : \tilde{\Omega} \to \mathbb{R}^2$ for the orthographic perspective setting. With n_i, $i = 1, 2, 3$, we refer to the components of a field of normal vectors, i.e., $\mathbf{n} = [n_1, \; n_2, \; n_3]^\top$. In the case of orthographic projection, the vector $g_{\perp}$ is constructed as

$$g_{\perp} = \left[-\frac{n_1}{n_3}, \; -\frac{n_2}{n_3} \right]^\top . \tag{10}$$

Solving the Poisson equation in (9) requires an adequately chosen boundary condition. Here, we utilize the so-called natural boundary condition $\langle \nabla u - g, \eta \rangle = 0$, with η being the normal vector to $\partial\tilde{\Omega}$ in the image plane Ω. The implementation of this boundary condition is not trivial, since $\tilde{\Omega}$ is in general not rectangular. For the technical details, we refer the interested reader to [4].

We note that (9) does not yield a unique solution, even with the natural boundary condition. This is because to any solution u^* we can add any constant $c \in \mathbb{R}$ such that $u^* + c$ is still a solution. This can be prevented by simply adding the term

$$\lambda \iint_{\tilde{\Omega}} \|u(x, y)\|^2 \, \mathrm{d}x\mathrm{d}y \tag{11}$$

to the minimized function in (9), with a small weight $\lambda > 0$, e.g., 10^{-9}. Thus, the resulting solution u^* will be around 0, which does not change its shape drastically.

In contrast to (2), in the perspective projection setting, the normal vector depends not only on the spatial derivatives of u, but also on the (positive) depth u itself. A method to circumvent this issue is to deploy an auxiliary variable $v := \ln u$. Thus, with focal length f and a carefully chosen normalization factor $1/u$, we obtain

$$\mathbf{n} = \frac{\tilde{\mathbf{n}}}{\|\tilde{\mathbf{n}}\|}, \qquad \tilde{\mathbf{n}} = \frac{1}{u}\begin{bmatrix} -f u_x \\ -f u_y \\ u + x u_x + y u_y \end{bmatrix} = \begin{bmatrix} -f v_x \\ -f v_y \\ 1 + x v_x + y v_y \end{bmatrix}.$$

Analogously to (9) and(10), the perspective case leads to

$$v = \underset{v}{\arg\min} \iint_{\tilde{\Omega}} \left\|\nabla v(x, y) - g_{\angle}(x, y)\right\|^2 \mathrm{d}x\mathrm{d}y \quad \Longleftrightarrow \quad \Delta v = \operatorname{div} g_{\angle},$$

with

$$g_{\angle}(x, y) = \begin{bmatrix} -n_1/(x n_1 + y n_2 + f n_3) \\ -n_2/(x n_1 + y n_2 + f n_3) \end{bmatrix}.$$

Again, a more detailed account can be found in [37].

Also in this case, natural boundary conditions and a regularizing term similar to (11) are employed. However, an additive constant for v translates into a multiplier for $u = \exp(v)$, which will usually lead to an incorrectly scaled reconstruction. To mitigate this, we employ the following heuristic. We compute two reconstructions $u_\perp$ and $u_\angle = \exp(v_\angle)$ while assuming orthographic and perspective projection, respectively. Then we simply compute the final reconstruction as $u^* = c_1 u_\angle$, where c_1 is the variable obtained through

$$(c_1, c_2)^\top = \underset{(c_1, c_2) \in \mathbb{R}^2}{\arg\min} \iint_{\tilde{\Omega}} \left\|u_\perp(x, y) - c_1 u_\angle(x, y) + c_2\right\|^2 \mathrm{d}x\mathrm{d}y.$$

The result is a perspective reconstruction with the multiplier chosen such that its shape is as close as possible to the orthographic reconstruction.

4 Hayakawa's Lighting Estimation Setup

As already pointed out in Sect. 2, we restructure the gray input image values as the $p \times q$ data matrix

$$\mathcal{M} = \begin{bmatrix}\mathbf{m}_1 \ \mathbf{m}_2 \cdots \mathbf{m}_q\end{bmatrix} = \begin{bmatrix} m_{11} & m_{12} & \cdots & m_{1q} \\ m_{21} & m_{22} & \cdots & m_{2q} \\ \vdots & \vdots & & \vdots \\ m_{p1} & m_{p2} & \cdots & m_{pq} \end{bmatrix},$$

where the matrix entry m_{kt} corresponds to the gray value of the kth pixel of the tth input image.

Let

$$\mathcal{R} = \begin{bmatrix} \rho_1 & & 0 \\ & \ddots & \\ 0 & & \rho_p \end{bmatrix},$$

be the surface reflectance diagonal matrix,

$$\mathcal{N} = \begin{bmatrix}\mathbf{n}_1 \ \mathbf{n}_2 \cdots \mathbf{n}_q\end{bmatrix} = \begin{bmatrix} n_{11} & n_{12} & \cdots & n_{1p} \\ n_{21} & n_{22} & \cdots & n_{2p} \\ n_{31} & n_{32} & \cdots & n_{3p} \end{bmatrix},$$

represent the $3 \times p$ surface normal matrix, and

$$\mathcal{L} = \begin{bmatrix}\boldsymbol{\ell}_1 \ \boldsymbol{\ell}_2 \cdots \boldsymbol{\ell}_q\end{bmatrix} = \begin{bmatrix} \ell_{11} & \ell_{12} & \cdots & \ell_{1q} \\ \ell_{21} & \ell_{22} & \cdots & \ell_{2q} \\ \ell_{31} & \ell_{32} & \cdots & \ell_{3q} \end{bmatrix},$$

the $3 \times q$ light source directions matrix. Then, we can write a discrete statement of Lambert's law (3) as the matrix equation

$$\mathcal{M} = \mathcal{R}\mathcal{N}^\top\mathcal{L}. \tag{12}$$

The PS technique under unknown lighting consists of computing the rank-3 factorization

$$\mathcal{M} = \widetilde{\mathcal{N}}^\top\mathcal{L}, \tag{13}$$

where $\widetilde{\mathcal{N}} = \mathcal{N}\mathcal{R}$ (cf. (12)), without knowing in advance the lights location, i.e., the matrix $\mathcal{L}$. Here, we briefly recall the method proposed by Hayakawa [21]; see also [15].

Let the compact singular value composition (SVD) [18] of the image data matrix be

$$\mathcal{M} = U\Sigma V^{\top}, \tag{14}$$

where the diagonal matrix

$$\Sigma = \text{diag}(\sigma_1, \ldots, \sigma_q) = \begin{bmatrix} \sigma_1 & & 0 \\ & \ddots & \\ 0 & & \sigma_q \end{bmatrix}$$

contains the singular values $\sigma_1 \geq \cdots \geq \sigma_q \geq 0$, and $U \in \mathbb{R}^{p \times q}$ and $V \in \mathbb{R}^{q \times q}$ are matrices with orthonormal columns $\mathbf{u}_i$ and $\mathbf{v}_i$. These are called the left and right singular vectors, respectively. As already remarked, in real PS applications, $q \ll p$.

When q is small, the SVD factorization can be computed efficiently by standard numerical libraries, even for a quite large value of p. In the case of a very large dataset, with the aim of reducing the computational complexity, a partial SVD may be constructed at a reduced cost [2, 3].

In (12), we assumed the data matrix $\mathcal{M}$ to have rank 3. In this situation,

$$\sigma_1 \geq \sigma_2 \geq \sigma_3 > \sigma_4 = \cdots = \sigma_q = 0.$$

Anyway, since images may be acquired in non-ideal conditions and may be affected by noise, factorization (14) usually has numerical rank $r > 3$. Then, a *truncated* SVD must be performed. This is achieved by adopting the partitioning

$$U = \begin{bmatrix} U_1 \ U_2 \end{bmatrix}, \qquad V = \begin{bmatrix} V_1 \ V_2 \end{bmatrix},$$

where U_1 and V_1 contain the first three columns of U and V, respectively, and letting $\Sigma_1 = \text{diag}(\sigma_1, \sigma_2, \sigma_3)$. Then, we consider the approximation

$$\mathcal{M} \simeq \mathcal{M}_1 = W^{\top} Z,$$

with $W = \Sigma_1 U_1^{\top} = [\mathbf{w}_1, \ldots, \mathbf{w}_p]$ and $Z = V_1^{\top} = [\mathbf{z}_1, \ldots, \mathbf{z}_q]$, which produces the best rank-3 approximation to the data matrix $\mathcal{M}$ in both the Euclidean and the Frobenius norm sense [8].

This initial rank-3 factorization is followed by the solution of the $q \times 6$ least-squares problem

$$\min_{\mathbf{g} \in \mathbb{R}^6} \|H\mathbf{g} - \mathbf{e}\|, \tag{15}$$

where $\mathbf{e} = (1, \ldots, 1)^\top \in \mathbb{R}^q$ and $H \in \mathbb{R}^{q \times 6}$ is the matrix whose t^{th} row is defined by

$$\left[z_{1t}^2 \; z_{2t}^2 \; z_{3t}^2 \; 2z_{1t}z_{2t} \; 2z_{1t}z_{3t} \; 2z_{2t}z_{3t}\right],$$

in terms of the elements of the columns $\mathbf{z}_t$ of Z.

The solution of the optimization problem (15) produces a vector $\mathbf{g}$ containing the entries in the upper triangle of a 3×3 symmetric positive definite matrix G, whose Cholesky factor R normalizes the columns of Z, in the sense that $\|R\mathbf{z}_t\| = 1$, $t = 1, \ldots, q$. The factors in the sought factorization (13) are given by

$$\widetilde{\mathcal{N}} = (R^{-1})^\top W, \qquad \mathcal{L} = RZ;$$

see [21] and [15]. Finally, the matrix $\mathcal{N}$ of the normal vectors is obtained by normalizing the columns of $\widetilde{\mathcal{N}}$, and the normalizing constants are the diagonal entries of the albedo matrix $\mathcal{R}$.

Hayakawa's procedure shows that the light positions can be detected only for $q \geq 6$, i.e., if at least 6 images with different lighting are available. Anyway, factorization (12) is unique up to a unitary transformation, and such a transformation has to be suitably chosen before proceeding with the normal integration, to ensure that the surface can be represented as an explicit function $z = u(x, y)$. A procedure for determining an acceptable surface orientation and to solve at the same time the so-called *bas-relief ambiguity* was proposed in [15, Section V].

5 The Oren–Nayar Model

The Oren–Nayar model [34] is designed to handle rough objects by modeling surfaces as an aggregation of many infinitesimally small Lambertian patches, called facets. A schematic view of a small portion $d\epsilon$ of a rough surface, made from a small set of facets, is displayed in Fig. 1.

The slope values of all such facets follow a Gaussian probability distribution with standard deviation $\sigma \in [0, +\infty)$, also called the *roughness parameter* of the surface. The main idea proposed in [34] is that each facet contributes to the modeled irradiance value $I_{\text{O-N}}$ according to the Oren–Nayar model, as follows:

$$I_{\text{O-N}} = \frac{\rho}{\pi} L_i \cos(\theta_i) \left(\nu_1 + \nu_2 \sin(\alpha) \tan(\beta) \max(0, \cos(\Phi_r - \Phi_i))\right), \qquad (16)$$

where (see also Fig. 2) ρ represents the facet albedo, L_i the intensity of the point-like light source, θ_i the angle between the surface normal and the light source, and θ_r the angle between the surface normal and the camera direction. In addition, two parameters $\alpha = \max(\theta_i, \theta_r)$ and $\beta = \min(\theta_i, \theta_r)$ represent the maximum and the

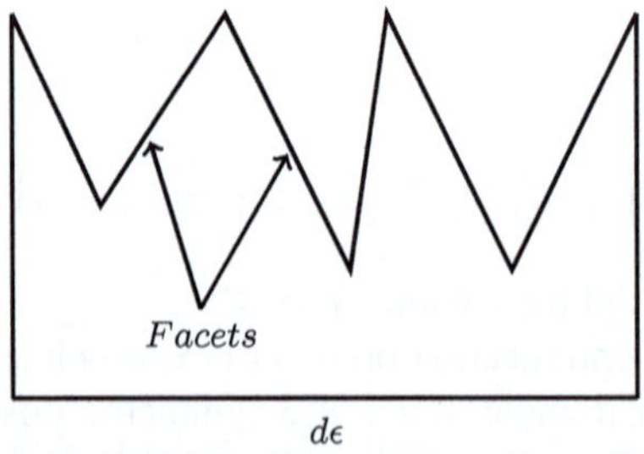

Fig. 1 A schematic side view of a small rough surface $d\epsilon$ formed by a few facets. The surface roughness is characterized by a Gaussian probability distribution of facet slopes with standard deviation $\sigma \in [0, +\infty)$. According to the Oren–Nayar model, each facet contributes to the irradiance value of the surface as shown in (16). Note that, in case of $\sigma = 0$, the surface follows the Lambertian model

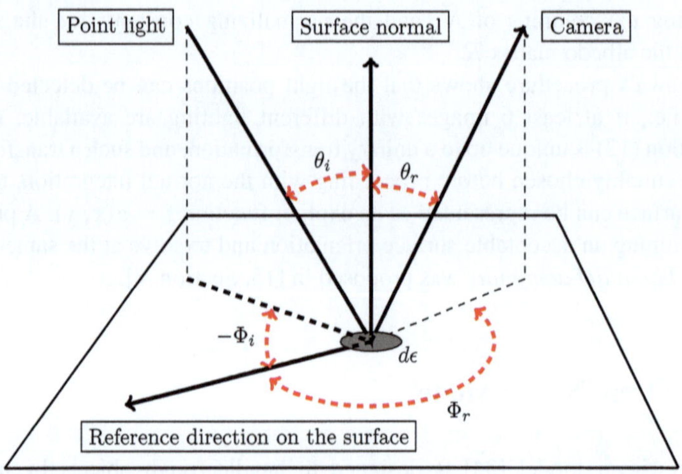

Fig. 2 Illustration of the Oren–Nayar model for the reflectance of a facet being illuminated by a point-like light source and captured by a camera. The directions from which the facet is observed and illuminated determine two angles θ_r and θ_i, respectively, with respect to the normal to the facet. In addition, the reference direction on the surface establishes two azimuth angles Φ_r and Φ_i for the camera and the illumination directions. Note that we do not visualize a particular facet because of its small size $d\epsilon$ compared to the surface area

minimum values of θ_i and θ_r, respectively, and the terms ν_1 and ν_2 depend on the roughness parameter σ

$$\nu_1 = 1 - 0.5\frac{\sigma^2}{\sigma^2 + 0.33} \qquad \text{and} \qquad \nu_2 = 0.45\frac{\sigma^2}{\sigma^2 + 0.09}.$$

Finally, Φ_i and Φ_r denote the azimuth angles for the light source and the camera direction, respectively, with respect to the reference direction on the surface, as shown in Fig. 2.

Next, we assume the point-like light source be located at the optical center of the camera and assume the constant coefficient $\frac{\rho}{\pi} L_i$ in (16) to be normalized to one. The second assumption is not restrictive, as it only depends on the light source intensity, the surface albedo, and the parameters of the imaging system, such as the lens diameter and the focal length; see [1]. This allows us to simplify (16) to

$$I_{\text{O-N}} = \nu_1 \cos(\theta) + \nu_2 \sin^2(\theta), \tag{17}$$

while the light source and viewing directions are considered to be coincident, resulting in $\theta_i = \theta_r = \alpha = \beta = \theta$ and $\Phi_i = \Phi_r$, with $\cos(\Phi_r - \Phi_i) = 1$. In the case of $\sigma = 0$, we have $\nu_1 = 1$ and $\nu_2 = 0$ in (17), and the Oren–Nayar model reduces to the Lambertian one.

A closer look at (17) reveals that the irradiance value $I_{\text{O-N}}$ based on the Oren–Nayar model consists of two components, namely, $\nu_1 \cos(\theta)$, the Lambertian one, and $\nu_2 \sin^2(\theta)$, which is the non-Lambertian component that attains its maximum when $\theta = \frac{\pi}{2}$.

Following the work by [39], we conclude the preprocessing phase by solving (17) for $\cos(\theta)$

$$\cos(\theta) = \frac{\nu_1 \pm \sqrt{\nu_1^2 - 4\nu_2 (I_{\text{O-N}} - \nu_2)}}{2\nu_2}, \tag{18}$$

and considering the solution associated to the minus sign, as motivated in [39]. Note that the obtained Lambertian component $\nu_1 \cos(\theta)$ based on (18) depends on the roughness parameter σ. This fact will be illustrated in our numerical results.

6 Numerical Results

Our dataset includes 20 images of a seashell illuminated from different directions by direct sunlight, three of which are shown in Fig. 3. To collect the images, a seashell with a width approximately equal to 10 (cm) was placed face up on a horizontal desk, with a tripod holding a camera about 100 (cm) above the seashell. The camera has a focal length of 85 (mm), and a black background was placed below the seashell to reproduce homogeneous Dirichlet boundary conditions for the observed surface. The desk was placed in the open air, under direct sunlight, and rotated in order to obtain 20 different lighting conditions; see Fig. 4. The Sun elevation angle was measured at the end of the shooting process; see [15] for a detailed explanation of the shooting procedure and also Fig. 5 for an overview on the modeling pipeline adopted in this chapter.

Fig. 3 Three out of 20 seashell images captured illuminated by the Sun from different directions, as elaborated in [15]

Fig. 4 Shooting setting for the seashell dataset. The whole system is placed upon a rotating platform

In addition, the maximum gray value corresponding to each pixel across all seashell images was determined, to apply a threshold to all of them. This establishes a mask based on a common contribution across all images. Finally, a morphological erosion [20] with a Euclidean disk of radius 3 pixels was used, providing the filtered seashell images for the next step. In this way, we may lose a few shell details on the boundary, but we ensure that there are no parts left in the interface actually belonging to the background. As an additional benefit, sharp peaks in the interface are smoothed.

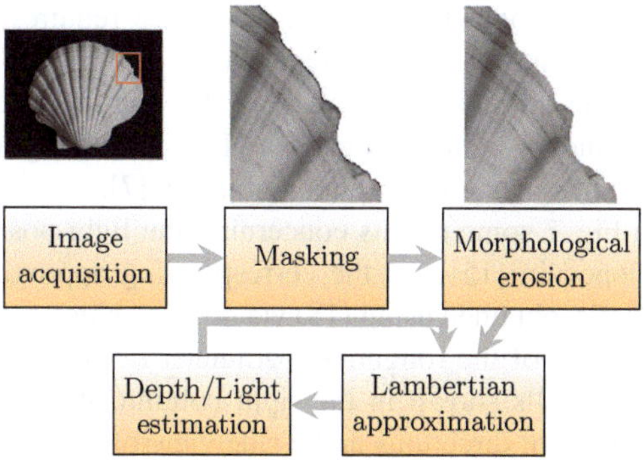

Fig. 5 Workflow adopted in the current work. We start by image acquisition, as explained in [21]. Next, masking and morphological erosion are used as preprocessing phases, so that the Lambertian components of the images can be approximated using the roughness value σ as a free parameter; see [34]. Finally, the optimal σ is found by using an optimization approach [10] that iteratively obtains the light directions and the best seashell model [15]. The optimization performed by [10] is represented by the loop between the last two steps, where σ is varied to produce a new Lambertian approximation, with the aim of producing the values nearest to the ground truth light directions acquired in the image acquisition step

Fig. 6 The filtered versions of the three images shown in Fig. 3

By considering each seashell to represent a rough surface, we adopt the Oren–Nayar model [34] and approximate the Lambertian component of the seashell surface, namely $\cos\theta$, by formula (18). After performing the computation, all values of $x = \cos\theta$ less than zero and greater than one, namely $-\kappa \leq x < 0$ and $1 < x \leq 1 + \kappa$ for a fixed κ, are mapped back to zero and one, respectively, to let the approximated values stay consistently in $[0, 1]$. These inaccuracies are introduced by the simplification assumed to derive (17) from (16), making the Oren–Nayar model adoptable in real-world applications [34]. In our dataset, the Lambertian approximation resulted in values between $[-\kappa, 1]$ with $\kappa = 0.3758$. Out of all approximated Lambertian values, 5.3% were smaller than zero, and none were larger than one.

In Fig. 6, we display a selection of the filtered images resulting from our preprocessing pipeline.

In the context of PS under unknown lighting conditions, we applied the procedure proposed in [21] to our masked seashell images to identify the lighting directions by the implementation from [15]. Such a procedure requires at least 6 images to estimate the light positions. As mathematically justified in [15], the resulting algorithm provides the possibility of inferring the lighting directions, leading to a PS problem with known light positions. In [15], a shooting technique was also introduced to solve the so-called bas-relief ambiguity [7].

We illustrate in Fig. 7 some results concerning the light positions identified by the software developed in [15] and the corresponding approximated Lambertian models of the seashells, obtained using (18) via orthographic projection for different values of σ. The impact of the roughness parameter σ can be clearly observed in the reconstructed seashells shown in the right column of Fig. 7. Here, each row contains the obtained light directions and a side view of the reconstructed seashell based on the approach proposed in [15]. The roughness parameter σ takes the values $\{1, 8, 25\}$ in degrees, from top to bottom. As it can be observed, larger values of σ cause the light direction to become less steep, and the seashell reconstructions to be more swollen.

On the basis of these results, we are motivated to determine a realistic range of variation for σ that results in a set of meaningful 3D reconstructions. In Fig. 8, we plot the residual sum of squares (RSS) between each Lambertian approximation and the corresponding original grayscale image of the seashell, for σ varying in $[\epsilon, 60]$, in degrees. Values larger than 60° may not be meaningful in practice; see [34]. The zero value is excluded since in this case the model directly downgrades to a Lambertian one based on (17) and (18) is not applicable any more. The curve shown in Fig. 8 is the mean of the curves corresponding to each of the 20 seashell images, constructed by varying $\sigma \in [\epsilon, 60]$ in steps of one degree. A close observation reveals that a value of σ close to 0° leads to a relatively high RSS value, revealing the dissimilarity between a grayscale image and its corresponding Lambertian approximation. This is expected, as for $\sigma = 0$ the Oren–Nayar model reduces to the Lambertian one. For $\sigma > 30°$, the curve starts fluctuating and clearly splits into two sub-curves. This motivates us to look for an optimal value of sigma in the range $[1°, 30°]$.

The 44.4° ground truth angle of the Sun above the horizon, which can be assumed to be constant for all images since the acquisition time was sufficiently small, is next used to determine σ. In practice, the optimal value proves to be 21.3795°, after letting σ vary in $[1°, 30°]$ with the aim of obtaining the smallest residual sum of squares (RSS) [17] between the elevation angles of the light directions estimated according to [15] and the ground truth value. This leads us to the inferred model (see Fig. 10) after 16 iterations, with a final RSS value of 171.6720.

The polar graph in Fig. 9 displays the optimal azimuth and elevation angles, estimated by the procedure in [15], for the optimal roughness parameter $\sigma = 21.3795°$. The θ- and r-directions correspond to the azimuth and elevation angles, respectively.

Finally, we report in Fig. 10 the 3D reconstruction based on the light directions represented in Fig. 9, assuming the seashell to be Lambertian with σ equal to 21.3795°. In particular, we display two views of the reconstructed object, a real picture of the shell, and a depth map of the 3D model.

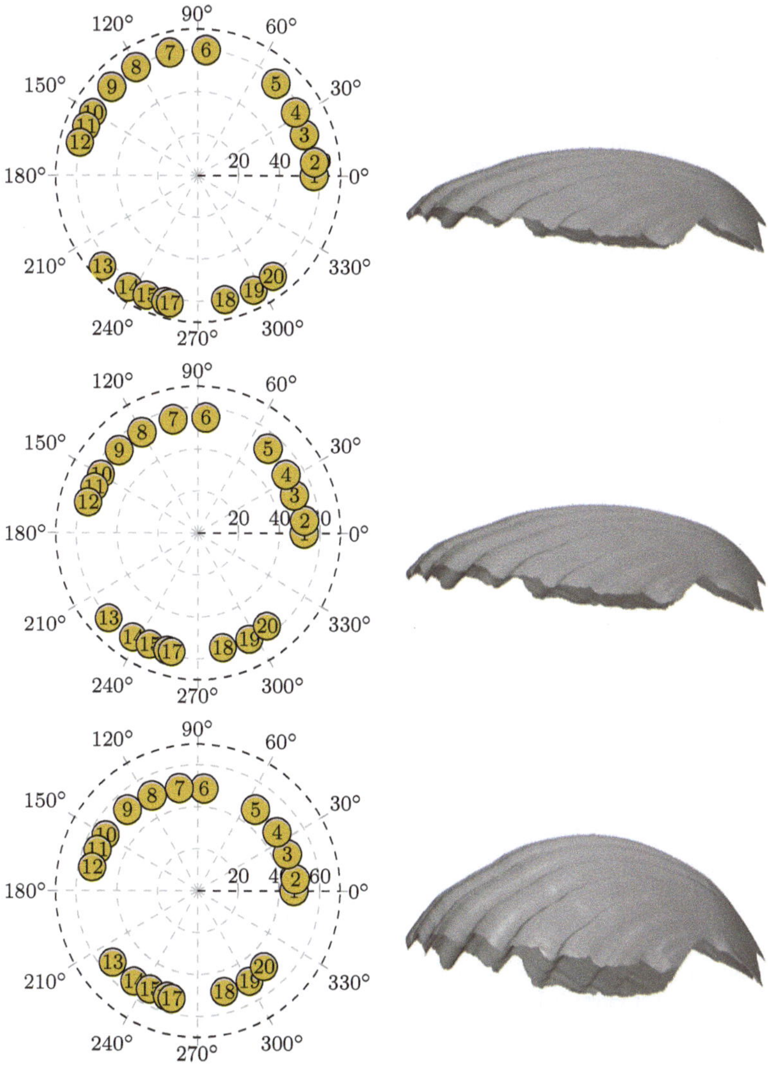

Fig. 7 The impact of the roughness parameter σ in estimating the light directions and in reconstructing the seashell by the method proposed in [15] can be clearly observed. Here, each row contains the results obtained for the roughness parameter $\sigma = 1°, 8°, 25°$, from top to bottom. A larger value of σ leads to a less steep estimate of the light direction. We observe in the right column a swelling effect on the reconstructed shapes, as a direct consequence of the increase in σ

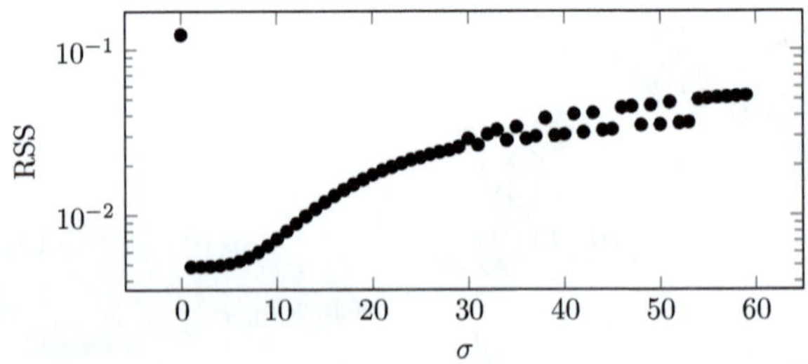

Fig. 8 Variation of the RSS between the Lambertian approximations and the corresponding original gray images of the seashell with values in [0, 1], for σ in $[\epsilon, 60°]$. Here, the RSS is used as a similarity measure. For $\sigma > 30°$, a fluctuation in the RSS is clearly observed, leading to two distinct branches of the curve. When σ is small, a large RSS reveals that the Lambertian modeling of gray images is meaningless. We conclude that a smooth variation of σ in $[1°, 30°]$ justifies the adoption of Oren–Nayar model, motivating us to consider this range as an educated initial guess to look for the optimal σ

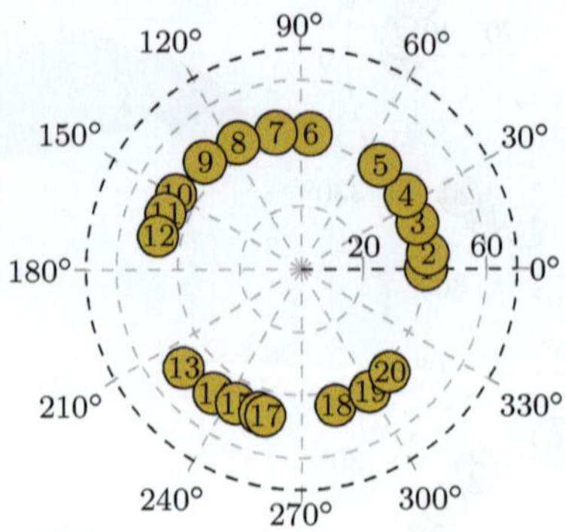

Fig. 9 Azimuth and elevation angles of the light directions estimated according to [15], when the Lambertian approximation corresponds to $\sigma = 21.3795°$. The azimuth angles, in the θ-direction, shed further light on the shooting procedure explained in [15]. Though we expect all the elevation angles, shown in the r-direction, to stay close to the measured ground truth angle 44.4°, in practice we obtain a range of values in $[40.21°, 49.34°]$

7 Summary and Conclusion

We developed a novel practical PS pipeline, and we rigorously motivated its constituent components. Our proposed model automatically estimates two fundamental parameters of the model, namely the unknown lighting environment and the reflectance properties of the observed object, whose unavailability often prevents PS to be robustly applied in real-world scenarios. Hayakawa's procedure detects the light positions, provided that at least 6 images of the sample object with different lighting directions are available. In addition, the reflectance properties of a rough object are approximated by the Oren–Nayar model to resemble a Lambertian surface. In practice, the Oren–Nayar model setting is used to tune the Hayakawa's

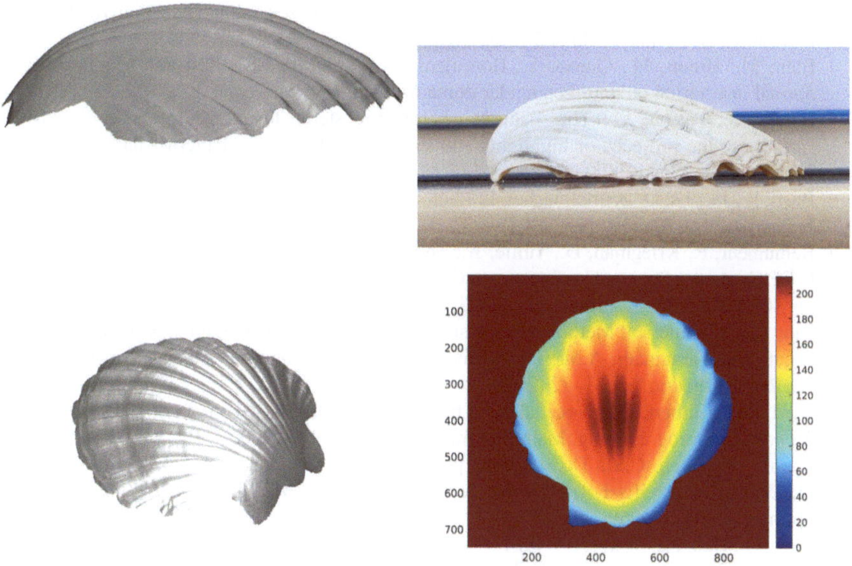

Fig. 10 The top rows display a side view of the reconstructed surface and a photo of the real shell from the same point of view. In the bottom row, we report another view of the reconstructed surface and the depth map of the 3D model of the seashell, by approximating it as a Lambertian surface based on the Oren–Nayar model [34]. The reconstruction is obtained using the light directions shown in Fig. 9

detected light directions, leading us to achieving the final optimally modeled object, as documented by our numerical results.

Acknowledgments The work of Georg Radow was supported by the Deutsche Forschungsgemeinschaft, grant number BR 2245/4-1.

The work of Giuseppe Rodriguez was partially supported by the Regione Autonoma della Sardegna research project "Algorithms and Models for Imaging Science [AMIS]" (RASSR57257, intervento finanziato con risorse FSC 2014-2020—Patto per lo Sviluppo della Regione Sardegna), and the INdAM-GNCS research project "Tecniche numeriche per l'analisi delle reti complesse e lo studio dei problemi inversi."

The work of Ashkan Mansouri Yarahmadi and Michael Breuß was partially supported by the European Regional Development Fund, EFRE 85037495.

References

1. Ahmed, A.H., Farag, A.A.: A new formulation for shape from shading for non-Lambertian surfaces. In: Proceedings of the 2006 IEEE Computer Society Conference on Computer Vision and Pattern Recognition (CVPR'06), vol. 2, pp. 17–22. IEEE, Piscataway (2006)
2. Baglama, J., Reichel, L.: Augmented implicitly restarted Lanczos bidiagonalization methods. SIAM J. Sci. Comput. **27**(1), 19–42 (2005)

3. Baglama, J., Reichel, L.: An implicitly restarted block Lanczos bidiagonalization method using Leja shifts. BIT Numer. Math. **53**, 285–310 (2012)
4. Bähr, M., Breuß, M., Quéau, Y., Boroujerdi, A.S., Durou, J.D.: Fast and accurate surface normal integration on non-rectangular domains. Comput. Vis. Media **3**(2), 107–129 (2017)
5. Barsky, S., Petrou, M.: The 4-source photometric stereo technique for three-dimensional surfaces in the presence of highlights and shadows. IEEE Trans. Pattern Anal. Mach. Intell. **25**(10), 1239–1252 (2003)
6. Basri, R., Jacobs, D., Kemelmacher, I.: Photometric stereo with general, unknown lighting. Int. J. Comput. Vis. **72**, 239–257 (2007)
7. Belhumeur, P., Kriegman, D., Yuille, A.: The bas-relief abiguity. In: Proceedings of the 1997 IEEE Computer Society Conference on Computer Vision and Pattern Recognition (CVPR'97), pp. 1060–1066. IEEE, Piscataway (1997)
8. Björck, A.: Numerical Methods for Least Squares Problems. SIAM, Philadelphia (1996)
9. Breuß, M., Yarahmadi, A.M.: Perspective shape from shading. In: Advances in Photometric 3D-Reconstruction. pp. 31–72 (2020)
10. Byrd, R.H., Gilbert, J.C., Nocedal, J.: A trust region method based on interior point techniques for nonlinear programming. Math. Program. **89**(1), 149–185 (2000)
11. Chen, C.P., Chen, C.S.: The 4-source photometric stereo under general unknown lighting. In: Leonardis, A., Bischof, H., Pinz, A. (eds.) Computer Vision – ECCV 2006, pp. 72–83. Springer, Berlin (2006)
12. Chen, G., Han, K., Wong, K.Y: PS-FCN: A flexible learning framework for photometric stereo. In: Proceedings of the European Conference on Computer Vision (ECCV), 16pp (2018)
13. Chen, G., Han, K., Shi, B., Matsushita, Y., Wong, K.: Self-calibrating deep photometric stereo networks. In: 2019 IEEE/CVF Conference on Computer Vision and Pattern Recognition, pp. 8731–8739 (2019)
14. Chen, G., Waechter, M., Shi, B., Wong, K.Y.K., Matsushita, Y.: What is learned in deep uncalibrated photometric stereo? In: European Conference on Computer Vision (2020)
15. Concas, A., Dessì, R., Fenu, C., Rodriguez, G., Vanzi, M.: Identifying the lights position in photometric stereo under unknown lighting. In: 2021 21st International Conference on Computational Science and Its Applications (ICCSA), pp. 10–20, Cagliari, Italy, September 2021
16. Durou, J.D., Falcone, M., Sagona, M.: Numerical methods for shape-from-shading: a new survey with benchmarks. Comput. Vis. Image Underst. **109**, 22–43 (2008)
17. Fox, J.: Applied Regression Analysis and Generalized Linear Models. SAGE Publishing, Thousand Oaks (2008)
18. Golub, G.H., Van Loan, C.F.: Matrix Computations. Johns Hopkins University Press, Baltimore (1996)
19. Haefner, B., Ye, Z., Gao, M., Wu, T., Quéau, Y., Cremers, D.: Variational uncalibrated photometric stereo under general lighting. In: Proceedings of the IEEE/CVF International Conference on Computer Vision, pp. 8539–8548 (2019)
20. Haralick, R.M., Sternberg, S.R., Zhuang, X.: Image analysis using mathematical morphology. IEEE Trans. Pattern Anal. Mach. Intell. **PAMI-9**(4), 532–550 (1987)
21. Hayakawa, H.: Photometric stereo under a light source with arbitrary motion. J. Opt. Soc. Am. A **11**(11), 3079–3089 (1994)
22. Hold-Geoffroy, Y., Zhang, J., Gotardo, P.F.U., Lalonde, J.F.: What is a good day for outdoor photometric stereo? In: International Conference on Computational Photography (2015)
23. Hold-Geoffroy, Y., Gotardo, P., Lalonde, J.F.: Single day outdoor photometric stereo. IEEE Trans. Pattern Anal. Mach. Intell. **43**(6), 2062–2074 (2021). https://doi.org/10.1109/tpami.2019.2962693
24. Horn, B.K.P. (ed.): Robot Vision. MIT Press, Cambridge, USA (1986)
25. Horn, B.: Obtaining shape from shading information. In: Shape from Shading, pp. 123–171 (1989)
26. Ikehata, S.: CNN-PS: CNN-based photometric stereo for general non-convex surfaces. In: Proceedings of the European Conference on Computer Vision (ECCV), 16pp (2018)

27. Ju, Y., Tozza, S., Breuß, M., Bruhn, A., Kleefeld, A.: Generalised perspective shape from shading with oren-nayar reflectance. In: Proceedings of the 24th British Machine Vision Conference, pp. 42.1–42.11. BMVA Press, Durham (2013)
28. Khanian, M., Boroujerdi, A.S., Breuß, M.: Photometric stereo for strong specular highlights. Comput. Vis. Media **4**, 83–102 (2018)
29. Kimmel, R., Siddiqi, K., Kimia, B.B., Bruckstein, A.M.: Shape from shading: level set propagation and viscosity solutions. Int. J. Comput. Vis. **16**(2), 107–133 (1995)
30. Kozera, R.: Existence and uniqueness in photometric stereo. Appl. Math. Comput. **44**(1), 1–103 (1991)
31. Lambert, J.H., Klett, M.J., Detlefsen, C.P.: Photometria Sive De Mensura Et Gradibus Luminis, Colorum Et Umbrae. Klett, Augustae Vindelicorum, Augustae Vindelicorum (1760)
32. Mecca, R., Falcone, M.: Uniqueness and approximation of a photometric shape-from-shading model. SIAM J. Imag. Sci. **6**(1), 616–659 (2013)
33. Mecca, R., Tankus, A., Wetzler, A., Bruckstein, A.M.: A direct differential approach to photometric stereo with perspective viewing. SIAM J. Imag. Sci. **7**(2), 579–612 (2014)
34. Oren, M., Nayar, S.K.: Generalization of the Lambertian model and implications for machine vision. Int. J. Comput. Vis. **14**(3), 227–251 (1995)
35. Phong, B.T.: Illumination for computer generated pictures. Commun. ACM **18**, 311–317 (1975)
36. Quéau, Y., Mecca, R., Durou, J.D., Descombes, X.: Photometric stereo with only two images: a theoretical study and numerical resolution. Image Vis. Comput. **57**, 175–191 (2017)
37. Quéau, Y., Durou, J.D., Aujol, J.F.: Normal integration: a survey. J. Math. Imaging Vision **60**(4), 576–593 (2017)
38. Radow, G., Hoeltgen, L., Quéau, Y., Breuß, M.: Optimisation of classic photometric stereo by non-convex variational minimisation. J. Math. Imaging Vision **61**(1), 84–105 (2019)
39. Ragheb, H., Hancock, E.R.: Surface radiance correction for shape from shading. Pattern Recogn. **38**(10), 1574–1595 (2005)
40. Santo, H., Samejima, M., Sugano, Y., Shi, B., Matsushita, Y.: Deep photometric stereo network. In: Proceedings of the IEEE International Conference on Computer Vision, pp. 501–509 (2017)
41. Shi, B., Mo, Z., Wu, Z., Duan, D., Yeung, S.K., Tan, P.: A benchmark dataset and evaluation for non-Lambertian and uncalibrated photometric stereo. IEEE Trans. Pattern Anal. Mach. Intell. **41**(2), 271–284 (2019)
42. Stocchino, G.: Mathematical Models and Numerical Algorithms for Photometric Stereo (Modelli Matematici e Algoritmi Numerici per la Photometric Stereo). Bachelor's Thesis in Mathematics, University of Cagliari (2015). Available at http://bugs.unica.it/gppe/did/tesi/15stocchino.pdf
43. Taniai, T., Maehara, T.: Neural inverse rendering for general reflectance photometric stereo (2018)
44. Tankus, A., Kiryati, N.: Photometric stereo under perspective projection. In: Proceedings of the IEEE International Conference on Computer Vision, pp. 611–616 (2005). https://doi.org/10.1109/ICCV.2005.190
45. Torrance, K.E., Sparrow, E.M.: Theory for off-specular reflection from roughened surfaces. J. Opt. Soc. Am. **57**(9), 1105–1114 (1967). http://www.osapublishing.org/abstract.cfm?URI=josa-57-9-1105
46. Tozza, S., Falcone, M.: A comparison of non-lambertian models for the shape-from-shading problem. In: Perspectives in Shape Analysis, pp. 15–42. Mathematics and Visualization, Springer (2016)
47. Tozza, S., Mecca, R., Duocastella, M., Bue, A.D.: Direct differential photometric stereo shape recovery of diffuse and specular surfaces. J. Math. Imaging Vision **56**(1), 57–76 (2016)
48. Vanzi, M., Mannu, C., Dessí, R., Tanda, G.: Photometric stereo for 3D mapping of carvings and relieves. Case studies on prehistorical art in Sardinia. In: Proceedings of Atti del XVII MAÇAO's Intemational Rock ArtSeminar (2016)
49. Woodham, R.J.: Photometric stereo: a reflectance map technique for determining surface orientation from image intensity. In: Optics & Photonics (1979)

50. Woodham, R.J.: Photometric method for determining surface orientation from multiple images. Opt. Eng. **19**(1) (1980)
51. Yi, J., Ni, H., Wen, Z., Liu, B., Tao, J.: CTC regularized model adaptation for improving LSTM RNN based multi-accent Mandarin speech recognition. In: 2016 10th International Symposium on Chinese Spoken Language Processing (ISCSLP), pp. 1–5 (2016)
52. Zhang, R., Tsai, P.S., Cryer, J., Shah, M.: Shape from shading: a survey. IEEE Trans. Pattern Anal. Mach. Intell. **21**, 690–706 (1999)

Shape-from-Template with Camera Focal Length Estimation

Toby Collins and Adrien Bartoli

Abstract One of the major and open research objectives in computer vision is to automatically reconstruct the 3D shape of a deformable object from a monocular image. Shape-from-template (SfT) methods use prior knowledge embodied in a template that provides the object's 3D shape in a known reference position, and a physical model that constrains deformation. SfT methods have shown great success in recent years; however, accurate methods require an intrinsically calibrated camera. This is an important practical limitation because the intrinsics of many real cameras are not available, so they must be estimated with a dedicated calibration process. In this chapter, we present a novel SfT method that handles unknown focal length (a critical intrinsic of the perspective camera). The other intrinsics such as the principal point and aspect ratio are assumed to take canonical values, which is valid for many real cameras. We call this problem fSfT, and we solve it by gradient-based optimization of a large-scale non-convex cost function. This is not trivial for two main reasons. First, it requires suitable initialization, and we present a multi-start approach using a small set of candidate focal lengths (typically fewer than three are required). We combine this with a mechanism to avoid repeated exploration of the search space from different starts. Furthermore, we present cost normalization strategies, allowing the same cost function weights to be used in a diverse range of cases. This is crucial to make the method practical for real-world use. The method has been evaluated on twelve public datasets, and it significantly outperforms a previous state-of-the-art fSfT method in both focal length and deformation accuracy.

Keywords Registration · Reconstruction · 3D model · Camera calibration · Deformation · Isometry

---

The research described in this chapter was conducted by Toby Collins while at the Institut Pascal as part of his Ph.D. thesis

---

T. Collins (✉) · A. Bartoli
Faculté de Médecine, Institut Pascal, Clermont-Ferrand, France

E. Cristiani et al. (eds.), *Mathematical Methods for Objects Reconstruction*,
Springer INdAM Series 54, https://doi.org/10.1007/978-981-99-0776-2_3

1 Introduction

1.1 Shape-from-Template (SfT)

Reconstructing the 3D shape of a deformable object from a monocular image is a central and open problem in computer vision. It is usually much harder compared to reconstructing rigid objects because of the significantly larger problem space and much weaker constraints. To make deformable reconstruction well-posed, prior knowledge is required. In many previous works, including this chapter, the prior knowledge is embodied in a *template* [4, 6, 36, 37, 42, 44, 58]. The template provides a textured 3D geometric model of the object in a reference shape (usually implemented as a surface mesh), and it also constrains how the object can physically deform from its reference shape. The template can be acquired by various means, for example, using a computer-assisted design (CAD) model, a 3D scanner, or using a reconstruction from monocular images viewing the object at rest with dense multi-view stereo (MVS). The approach of solving monocular deformable reconstruction with a template is often called *shape-from-template* (SfT) in the literature, or equivalently *template-based monocular deformable reconstruction*. We use SfT in this chapter.

SfT is ill-posed if the template can deform arbitrarily because of the loss of depth information resulting from camera projection. To overcome this, most methods use a quasi-isometric template that prevents deformations that significantly stretch or shrink the template. This deformation model is valid for many objects of interest including those made of leather, plastic, stiff rubber, paper and cardboard, and tightly woven fabrics. Crucially, a quasi-isometric template can guarantee that SfT is well-posed with a calibrated perspective camera [4, 42]. SfT has various applications that include augmented reality (AR) with deformable objects, human–computer interaction (HCI) with deforming objects, and AR-guided surgery [12, 36, 54]. We illustrate two of these applications in Figs. 1 and 2.

Fig. 1 An HCI and AR application of SfT from [30]. This is an interactive educational game, implemented on a tablet, for coloring a virtual 3D cartoon model using a real coloring book. A page from the book is colored with a pencil, and SfT is used to register images of the page with a paper template. The registration from SfT allows the color from the image to be transferred to the paper template. Using a known association between the template and the cartoon model, the color is then transferred to the cartoon model. Because SfT also provides the 3D deformation of the paper template, the cartoon model can be virtually positioned on the paper sheet in real time

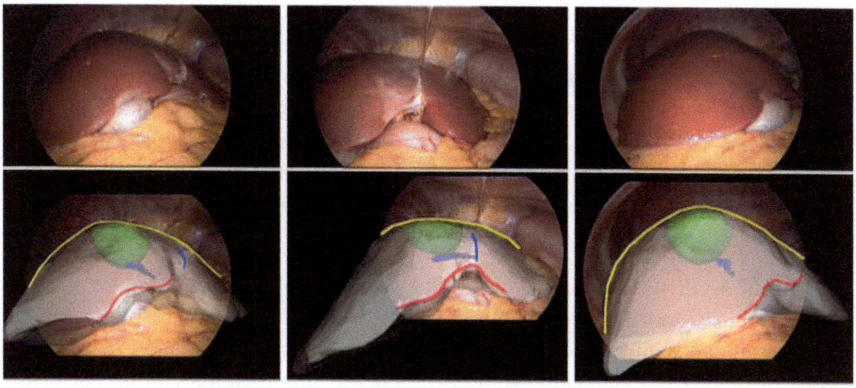

Fig. 2 An AR application of SfT from [25]. This is a prototype system to assist laparoscopic surgery of the liver using AR guidance, for safer liver resection. The top row of images shows three frames from a laparoscopic video of a liver. The bottom row shows the images augmented with hidden anatomical structures, including a tumor shown in green. This has been achieved using SfT, with a template constructed from a pre-operative CT image of the liver. The template is registered with laparoscopic images using SfT with contour and shading constraints

1.2 Chapter Innovations

SfT has been studied extensively with a perspective camera that is fully calibrated [4, 6, 36, 37, 42, 44, 58]. Calibrated intrinsics are required to relate camera coordinates with image coordinates. However, requiring known intrinsics is an important limitation in many real-world applications. A camera may have fixed and unknown intrinsics, or time-varying and unknown intrinsics, e.g., if the camera zooms in or out. Neither situation can be handled by these methods. With fixed and unknown intrinsics, a classical camera calibration is usually performed with a rigid calibration target such as a checkerboard [59]. However, this has several limitations. The calibration process requires user interaction and time, it adds inconvenience to the user, and a calibration target may not be available. Furthermore, *a priori* calibration is only suitable when the camera intrinsics are fixed, which is restrictive.

This work describes a novel SfT algorithm that jointly estimates focal length and the template's 3D deformation from a single image. We refer to this problem as *focal length and shape-from-template* (fSfT). The other intrinsics are assumed to take canonical values. fSfT is an important problem because many real cameras can be modeled accurately with negligible skew, an aspect ratio of one, a principal point at the image center, and negligible lens distortion. The only unknown intrinsic is the focal length. Our approach also works if the non-focal length intrinsics have known non-canonical values, computed with a calibration process; however, this is a less common use case.

We solve fSfT by designing and optimizing a large-scale non-convex cost function $c(f, \boldsymbol{\theta})$, where f is the unknown focal length and $\boldsymbol{\theta}$ is the unknown template deformation. The form of c is similar to cost functions used by the most accurate SfT methods that require calibrated cameras [12, 36]. The cost function includes a data cost to register the template and a deformation cost to penalize non-isometric deformation. We cannot optimize c with guaranteed global optimality. Nevertheless, we present a solution that works very well in practice, using local (iterative) optimization, combining a well-designed initialization strategy, careful cost modeling, and fast optimization. The principal novel characteristics and advantages of the approach are as follows:

1. We model *all* deformation constraints provided by the template in the cost function using a mesh-based physical deformation model. The results we obtain are significantly more accurate compared to the analytical method [3].
2. Precise initialization is not required in general. Initialization can be performed either using the analytical fSfT method [3] or using a very small number of focal length samples (three or fewer). We introduce a mechanism to improve computational efficiency to avoid repeated optimization in the same region of search space from different initializations.
3. We apply normalization techniques to the cost function, which greatly reduces the need to tune cost weights. Such tuning is a known issue in cost optimization approaches, and thanks to normalization, the same weights can be used for any problem instance. In our experimental evaluation, the same weights are used in *all* test cases, covering different object shapes, mesh discretization, textures, deformations, and imaging conditions. The ability to use the same weights in all conditions represents a significant advance toward a practical SfT solution.

1.3 Chapter Organization

The remainder of this chapter is organized as follows. In Sect. 2, we summarize previous approaches for solving SfT and fSfT, and we discuss their main limitations. In Sect. 3, we describe our fSfT method in detail. In Sect. 4, we present the experimental results, and in Sect. 5, we present our conclusion and directions for future research.

2 Related Works

2.1 SfT Approaches

We categorize prior SfT methods into three main groups: (*i*) closed-form methods that do not require an initialization, (*ii*) optimization-based methods that are generally more accurate than closed-form methods but require an initialization, and more recently (*iii*) convolutional neural network (CNN)-based methods. We now review these three categories.

2.1.1 Closed-Form Solutions

There are two main ways to solve SfT in closed form. The first relaxes the isometric constraint to inextensibility [5, 37, 42], which allows the surface to shrink but not stretch. The problem is then cast as finding the deformation that maximizes the depth of matched points such that the Euclidean distances between surface points do not exceed their geodesic distances (as defined by the template). The problem is convex and has been solved using a greedy approach [42] and with second-order cone programming (SOCP) using the interior point method [37, 42]. When the perspective effects are strong and there are many points, these methods can be very accurate. However, performance deteriorates when perspective effects and/or the number of points are reduced [8].

The second main way to solve SfT in closed form uses 1st-order non-holonomic partial differential equations (PDEs) [4]. A PDE is set up at each surface point that relates surface depth, normals, camera projection, and registration functions to 1st order. By imposing the isometric constraint, the PDE can be solved analytically by treating depth and normals as independent functions (a problem relaxation). The approach can also solve conformal (angle preserving) deformation [4] up to an arbitrary scale factor and convex/concave ambiguities. The PDE approach is very fast, and it can be parallelized trivially. However, an accurate registration is normally required, which can be hard with poorly textured surfaces.

2.1.2 Optimization-Based Solutions

The main disadvantage of the closed-form methods is to relax physical constraints, yielding sub-optimal solutions. In contrast, optimization-based solutions can exploit all available physical constraints. They take as input an initial sub-optimal solution and perform iterative numerical optimization of a non-convex cost function [12, 28, 31, 36, 44, 58, 58]. Practically, all methods use a pseudo-*maximum a posteriori* cost function consisting of prior and data terms. The prior term nearly always penalizes non-isometric deformation. The data term penalizes disagreement between the deformed template and image evidence, such as the reprojection of point matches [4, 8, 37, 42], patch-based matches [12], pixel-level photo consistency [31, 35, 58], or contours [13, 17, 22, 55]. The main advantage of optimization-based methods is that complex cost functions can be used with no known closed-form solution. When properly initialized, they generally produce the most accurate solutions. Initialization can be performed using a closed-form method, or in the case of video data, with the solution from the previous frame (also called *frame-to-frame tracking*). There are three main open challenges with optimization-based solutions. The first is to increase the convergence basin, to reduce the dependency on good initialization. Methods such as coarse-to-fine optimization with multi-resolution meshes [58] or advanced schemes using geometric multi-grid [12] have proved useful. The second challenge is to reduce the cost of optimization for real-time solutions. This has been achieved with dimensionality reduction and GPU implementations such as [12]. The third challenge is designing a cost function that works well in a broad range of settings without requiring fine-tuning of hyper-parameters.

2.1.3 CNN-Based Solutions

CNNs have been used with great success for solving monocular reconstruction problems with deformable objects, such as 3D human pose estimation [21, 32], surface normal reconstruction [1, 56], and monocular depth estimation [14, 19, 27]. These works have stimulated recent progress for solving SfT with CNNs [15, 16, 20, 40]. The main idea is to train a CNN to learn the function that maps a single RGB image with known camera intrinsics to the template's deformation parameters. The CNNs in these works are trained using supervised learning with labeled data, i.e., pairs of RGB images with the corresponding deformation parameters. Acquiring labeled data is a main practical challenge, and it is practically impossible to obtain with real data. For this reason, these works rely heavily on simulated labeled data generated by rendering software such as Blender. On one hand, this offers a way to generate an enormous amount of training data. On the other hand, this opens up new challenges to ensure that the training data represent the variability and realism of real-world images. The so-called render gap is a term used to express the difference in realism between simulated and real data, and it affects the ability of the CNN to generalize well to real data. In SfT, we additionally face the problem that the space of possible deformations can be exceptionally large, making it difficult to cover the deformation space sufficiently with training data. For this reason, these works have been shown to work with objects undergoing simple, smooth deformation with a low-dimensional deformation space such as bending paper sheets or smoothly deforming cloth. Furthermore, these works require intrinsically calibrated cameras. There has been some recent progress for combining labeled simulated data with partially labeled real data in order to reduce the render gap [15]. The real data are acquired by a standard RGBD camera. These data do not contain sufficient information to train the CNN with supervised learning because RGBD images provide depth but not registration information. Consequently, the CNN is trained with a combination of supervised learning (to learn the template's depth) and unsupervised learning (to learn the template's registration). Unsupervised learning is implemented using a photometric loss similar to multi-scale normalized cross-correlation.

While [15] mark a good step forward to solving SfT with CNNs, it requires calibrated RGBD data, so it is not applicable for solving fSfT. Furthermore, it requires a CNN to be trained specifically for each template, which is a strong practical and computational limitation. This directly contrasts our approach to fSfT, which does not require a computationally intensive training process for each template, making it much easier to apply in real applications. Very recently, a CNN-based approach has been presented that eliminates the need to train for a specific template texture [16]. This is promising work; however, it only works for flat, rectangular surfaces such as a sheet of paper. This contrasts our approach to fSfT that handles templates with any shape or texture.

2.2 *fSfT Solutions*

There have been a few previous approaches to solve fSfT [2, 3]. An approach using affine correspondences (ACs) [33] has been presented using focal length sampling [2]. Given a focal length sample, the depth of each AC can be estimated using the AC's motion with plane-based pose estimation [11, 23, 49, 59]. A good focal length sample should produce reconstructions that satisfy the isometric assumption (specifically that the Euclidean distance between reconstructed neighboring ACs is similar to their geodesic distance that is known *a priori* from the template). The method densely samples candidate focal lengths, and for each candidate, reconstruction compatibility is tested with the isometric assumption. However, this approach has several shortcomings. First, it requires precisely registered ACs, which are difficult to achieve in practice, and it normally requires iterative registration refinement that is computationally expensive. Second, it cannot compute focal length in closed form. Third, it was only shown to work well with strongly textured surfaces with many ACs.

An improved fSfT method that estimates focal length analytically has also been presented [3]. Using point correspondences, a local smooth warp is fitted to point neighborhoods, and focal length is then estimated at each point using a 2nd-order PDE. In a final step, focal length estimates are robustly combined from multiple neighborhoods. This approach is fast and works well for smooth, well-textured surfaces. However, it is sub-optimal because it does not apply geometric constraints acting across point neighborhoods. These are essential to obtain accurate solutions especially when point correspondences are sparse.

fSfT has similarities with the problem of texton-based shape-from-texture with focal length estimation [10]. In texton-based shape-from-texture, the goal is to reconstruct a surface whose texture consists of repeated units known as textons. If the textons are small, such as the circular dots on a polka dot dress, they can each be modeled well by a plane. When the texton's metric shape is known *a priori* (for example, knowing that the textons are circular), each texton can be reconstructed with plane-based pose estimation. The whole surface can then be reconstructed by interpolation or surface normal integration. [10] propose two analytical approaches to texton-based shape-from-texture with unknown focal length. The first solves f using the fact that the Euclidean distances between neighboring textons are approximately preserved by isometric deformation, producing a unique solution to f with a minimum of two textons. The second finds f that yields an integrable surface. There are similarities between [10] and the fSFT solutions, where each texton can be considered as a local template or a single AC. Additionally, both [10] and [3] use a weak-perspective approximation to obtain an analytical solution. The main limitation of [10] is to only solve texton reconstruction, and it cannot handle general objects or textures unlike the proposed fSfT method.

3 Methodology

3.1 Problem Modeling

3.1.1 Template Geometry and Deformation Parameterization

The setup is illustrated in Fig. 3. The template surface $\mathcal{R} \subset \mathbb{R}^3$, defined in object coordinates, is modeled using a discrete texture-mapped triangulated surface mesh, called the *template mesh*. The template mesh has connected and non-overlapping triangle faces that model the surface piecewise linearly. It consists of vertices $\mathcal{V}$, edges $\mathcal{E}$, and faces $\mathcal{F}$. We define as $\boldsymbol{y}_{i\in[1,V]} \in \mathcal{R}$ the known 3D position of vertex i in object coordinates, where V is the number of vertices. We define as $\boldsymbol{\theta}_{\text{ref}}$ the known 3D positions of all vertices in object coordinates, corresponding to the template's reference shape:

$$\boldsymbol{\theta}_{\text{ref}} \stackrel{\text{def}}{=} \text{stk}\left(\boldsymbol{y}_1, \boldsymbol{y}_2, \ldots, \boldsymbol{y}_V\right) \tag{1}$$

where stk is the stacking operator that concatenates its arguments into a column vector. We define as $\boldsymbol{x}_{i\in[1,V]} \in \mathbb{R}^3$ the unknown 3D position of vertex i in camera

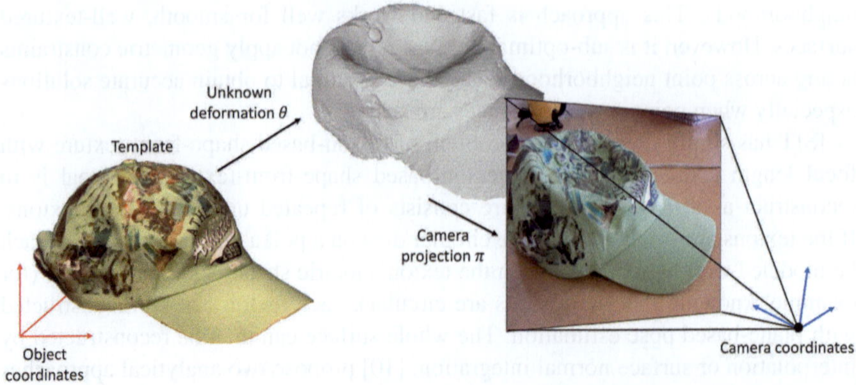

Fig. 3 SfT illustrated with a deformable cap [2]. The goal of SfT is to determine the unknown deformable 3D transform $\boldsymbol{\theta}$ that maps the template to camera coordinates, using (i) visual information in the image and (ii) deformation prior knowledge embodied in the template. The goal of fSfT is to solve SfT and jointly calibrate the camera's unknown focal length

coordinates, and we define as $\boldsymbol{\theta}$ the unknown positions of all vertices in camera coordinates:

$$\boldsymbol{\theta} \stackrel{\text{def}}{=} \text{stk}\left(\boldsymbol{x}_1, \boldsymbol{x}_2, \ldots, \boldsymbol{x}_V\right). \tag{2}$$

We define as $g(\boldsymbol{p}; \boldsymbol{\theta}) : \mathcal{R} \rightarrow \mathbb{R}^3$ the spatial transformation of a surface point $\boldsymbol{p} \in \mathcal{R}$ from object coordinates to camera coordinates. This is parameterized using $\boldsymbol{\theta}$ with barycentric interpolation. Specifically, $\boldsymbol{p}$ is uniquely associated to an enclosing mesh triangle and transformed according to the motion of the triangle's three vertices:

$$g(\boldsymbol{p}; \boldsymbol{\theta}) : \mathcal{R} \rightarrow \mathbb{R}^3 \stackrel{\text{def}}{=} w_1 \boldsymbol{x}_i + w_2 \boldsymbol{x}_j + w_3 \boldsymbol{x}_k, \tag{3}$$

where i, j, and k denote the three indices of the enclosing triangle, and $0 \leq w_1, w_2, w_3 \leq 1$ are the known barycentric weights associated with point $\boldsymbol{p}$ such that $w_1 + w_2 + w_3 = 1$.

3.1.2 Cost Function

General Form

We model fSfT with a non-convex cost function $c(\boldsymbol{\theta}, f) : \Theta \times \mathbb{R}^+ \rightarrow \mathbb{R}^+$ that maps deformation parameters $\boldsymbol{\theta}$ and focal length f to a positive real cost. We recall that $\boldsymbol{\theta}$ has been defined in Eq. 2 as the unknown 3D positions of the template's vertices in camera coordinates.

We use a cost function inspired from the SfT literature with special attention to cost normalization to ensure it works well for a broad variety of problem instances (templates, deformations, viewpoints, textures, *etc.*). The cost function is a weighted combination of three terms as follows:

$$c(\boldsymbol{\theta}, f; \mathcal{P}, \mathcal{Q}) = c_{\text{data}}(\boldsymbol{\theta}, f; \mathcal{P}, \mathcal{Q}) + \lambda_{\text{iso}} c_{\text{iso}}(\boldsymbol{\theta}) + \lambda_{\text{reg}} c_{\text{reg}}(\boldsymbol{\theta}). \tag{4}$$

The terms c_{data}, c_{iso}, and c_{reg} are the data, isometric, and regularization costs, respectively. The terms λ_{iso} and λ_{reg} are weights that balance the influence of c_{iso} and c_{reg} and are important hyper-parameters. Note that f influences only c_{data} directly and it influences the other terms indirectly via $\boldsymbol{\theta}$. In contrast, $\boldsymbol{\theta}$ influences all terms directly. As a pre-processing step, we normalize the template's size by a scale factor s so its total surface area is 1 unit. This makes the cost function invariant to templates of different sizes. After optimization, the deformed template is recovered in its original size by scaling the solution to $\boldsymbol{\theta}$ by $\frac{1}{s}$. We now summarize our implementations of each term.

Data Cost

The data cost c_{data} forces registration between the template's surface and the image from point correspondences. We denote the correspondences by the ordered sets

$\mathcal{P} \in \mathbb{R}^{3N}$ and $\mathcal{Q} \in \mathbb{R}^{2N}$, where each point $\mathcal{P}(i \in [1, N]) \in \mathbb{R}^3$ is a point on the template surface in object coordinates, and $\mathcal{Q}(i \in [1, N]) \in \mathbb{R}^2$ is the point in image coordinates. Each point correspondence is related as follows:

$$\pi(g(\mathcal{P}(i), \boldsymbol{\theta}); f) = \mathcal{Q}(i) + \epsilon_i, \tag{5}$$

where $\epsilon_i \in \mathbb{R}^2$ is unknown measurement noise and $\pi : \mathbb{R}^3 \to \mathbb{R}^2$ is the perspective projection function that depends on the unknown focal length f:

$$\pi\left(\begin{pmatrix} x \\ y \\ z \end{pmatrix}; f\right) \stackrel{\text{def}}{=} \frac{f}{z}\begin{pmatrix} x \\ y \end{pmatrix}. \tag{6}$$

Various approaches can be used to determine point correspondences, and our method is not tied to a specific approach. One of the most common approaches used in previous SfT methods is image keypoint matching (also known as interest point matching), where standard methods such as SIFT [29] or learning-based methods such as LIFT [57] may be used. First, keypoints are detected in one or more images of the template, which are then back projected onto the template's surface to determine their barycentric coordinates. A second set of keypoints are then detected in the input image, and the two keypoint sets are matched based on keypoint descriptors to generate $\mathcal{P}$ and $\mathcal{Q}$. Often keypoint matching methods generate mismatches, which are point correspondences that do not physically correspond to the same surface point up to noise. In a pre-processing step, we remove mismatches with a dedicated method. Mature methods exist that can be used without knowledge of camera intrinsics. Possible methods include [36] where the template is fitted directly in 2D using a stiff-to-flexible annealing scheme, RANSAC-based model fitting such as [52], or methods based on motion consistency between neighboring points [39]. Our approach can be used with any combination of point matching and outlier detection methods, and we give our implementation choices for these in the experimental section of this chapter.

Outlier rejection methods are not always perfect, and our SfT method includes robustness built-into c_{data} to handle a small proportion of residual mismatches. This is implemented with the Huber M-estimator ρ_h, and the data cost writes as follows:

$$\begin{array}{ll} c_{\text{data}}(\boldsymbol{\theta}, f; \mathcal{P}, \mathcal{Q}) = \frac{1}{N}\sum_{i=1}^{N}\frac{1}{\sigma^2}\rho\left(\pi(g(\mathcal{P}(i); \boldsymbol{\theta}), f) - \mathcal{Q}(i)\right) & (a) \\ \rho(\text{stk}(x, y)) \stackrel{\text{def}}{=} \rho_h(x) + \rho_h(y) & (b) \\ \rho_h(z) \stackrel{\text{def}}{=} \begin{cases} \frac{1}{2}z^2 & \text{if } |z| < k \\ k(|z| - \frac{1}{2}k) & \text{otherwise} \end{cases} & (c) \end{array}. \tag{7}$$

The M-estimator acts to reduce the influence of correspondences with large residuals, which are commonly caused by mismatched points. The term σ is an estimate of the noise standard deviation. Its value depends on several factors, including the method used to generate point correspondences, image resolution, and image noise. Unless σ is known, we use the following as default:

$$\sigma = \frac{1}{640}\max(w, h). \tag{8}$$

The image resolution (w, h) is taken into account in Eq. (8) as σ is scaled by $\max(w, h)$. The denominator (640) is merely intended to help interpret σ relative to VGA resolution. The default defined in Eq. (8) corresponds to a noise standard deviation of 1 pixel at VGA resolution. The value k is the Huber constant, set to a default $k = 10\,\sigma$.

Isometric Cost

The isometric cost is implemented using a discrete approximation of the elastic strain energy E_{strain} of continuous surfaces [50]:

$$E_{strain} = \int_{\mathcal{R}} \|\mathrm{I}_{\mathcal{R}} - \mathrm{I}_{\mathcal{S}}\|_F^2 \, \mathrm{d}\mathcal{R}, \tag{9}$$

where $\mathrm{I}_{\mathcal{R}}$ and $\mathrm{I}_{\mathcal{S}}$ are the first fundamental forms of $\mathcal{R}$ (the template's surface in object coordinates) and $\mathcal{S}$ (the template's surface in camera coordinates), respectively. Penalizing E_{strain} encourages a deformation to preserve the first fundamental form, equivalent to penalizing non-isometric deformation. We use a discrete approximation of E_{strain} using a finite element model (FEM) with constant strain triangles (CSTs). This is a well-known model from mechanics that is suitable for relatively stiff (quasi-isometric) materials. Furthermore, using a FEM with CSTs gives a consistent discretization of the continuous strain energy. That is, under appropriate refinement conditions and norms, it is largely invariant to the mesh discretization, and it converges to the continuous energy E_{strain}. This is important for our purposes because it eliminates the need to tune the cost's weight λ_{iso} according to the mesh discretization (the number of vertices, placement of vertices, and the triangulation). This is not true for the majority of membrane-like costs used in the SfT, which often use inconsistent isometric costs, such as those based on the preservation of mesh edge lengths [5, 17, 36] or the popular as-rigid-as-possible (ARAP) cost from [47]. The ARAP cost was shown to not be a consistent scheme in [26].

We compute the isometric cost c_{iso} as the discrete approximation of Eq. (9) using CSTs. This is implemented by a weighted sum of strain energies from each triangle, and the implementation details are given in Appendix 2. The isometric cost is a cubic expression in $\boldsymbol{\theta}$, which makes the minimization of c a large-scale non-convex problem.

Regularization Cost

The regularization cost c_{reg} is convex, and it encourages smooth deformation. Various implementations could be used, and we use a simple one using the moving least-squares energy [46], also used in [12]. First, the mesh is divided into overlapping *cells* where each cell describes the local motion of the mesh. We define one cell per vertex, containing all neighboring vertices connected by a mesh edge. The cell's motion is determined by the movement of its constituent vertices. Regularization is imposed by encouraging the cell's motion from object to camera coordinates to be described with an affine transform that is specific to each cell. This is implemented by penalizing the residual of the least-squares affine motion of each cell. It is straightforward to show that the residuals are linear in $\boldsymbol{\theta}$, making c_{reg} convex and quadratic in $\boldsymbol{\theta}$. However, unlike c_{iso}, c_{reg} is not consistent, which means it depends strongly on the mesh discretization. The reason is similar for why the ARAP mesh energy is not consistent as discussed in [26]. Indeed, constructing a consistent and convex regularization cost with surface meshes is not trivial, and it has not been achieved before in the SfT literature. We handle this using normalization as follows. We apply a global reweighing to c_{reg} so that a small deformation from the rest state induces approximately the same cost irrespective of the template's discretization:

$$c_{\text{reg}} \leftarrow \frac{1}{\left\| \boldsymbol{J}_{\text{reg}} \right\|_F^2} c_{\text{reg}}, \tag{10}$$

where $\boldsymbol{J}_{\text{reg}}$ is the Jacobian matrix of c_{reg}.

3.1.3 Cost Normalization Summary and Weight Hyper-parameters

In the cost definitions above, normalization has been used to significantly reduce the need to tune the cost weight hyper-parameters λ_{iso} and λ_{reg}. Specifically, normalization has made c strongly invariant to four sources of variability: template scale, template discretization, the number of correspondences, and image resolution. The influence of template scale is handled by rescaling the template to have unit area. The influence of template discretization is handled by two techniques. The first technique, used to normalize c_{iso}, involves the use of a consistent discrete approximation of a continuous surface function (strain energy) with an FEM, which achieves good discretization invariance by construction. The second technique, used to normalize c_{reg}, involves re-weighting the cost term by the magnitude of its Jacobian. This results in a small deformation from the rest state having approximately the same regularization cost irrespective of the discretization.

Invariance to the number of correspondences N is achieved by rescaling c_{data} inversely in N in Eq. (8). Image resolution invariance is achieved in Eq. (8) by rescaling the residual error of each point inversely by the image size. Note that image resolution invariance is often achieved in SfT methods by defining residual errors in retinal coordinates (also called normalized pixel coordinates). However, this does not work for fSfT because it yields a trivial solution with the focal length at infinity.

Thanks to these normalization techniques, we use the same weights λ_{iso} and λ_{reg} for *all* templates and test datasets, where mesh resolutions vary considerably from $O(100)$ to $O(1000)$ vertices, the number of point correspondences varies from $O(10)$ to $O(1000)$, and image resolutions vary from VGA to high definition (3600×2400 pixels). In all experiments, we use a default of $\lambda_{\text{iso}} = 1583$ and $\lambda_{\text{reg}} = 1\text{e} - 3$, found experimentally.

3.2 Optimization

3.2.1 Approach Overview

The cost function c defined in Eq. (4) is non-convex in the unknowns f and $\boldsymbol{\theta}$, arising from the non-convexity of both c_{iso} and c_{data}. Concerning c_{data}, the non-convexity is from the depth division of π. Concerning c_{iso}, the non-convexity is because c_{iso} is quartic in $\boldsymbol{\theta}$ as detailed in Sect. 2.2 of the appendix. Although f appears only directly in c_{data}, it has an indirect influence on c_{data} and c_{iso} by its connection to $\boldsymbol{\theta}$ in c_{data}.

Our goal is to determine f and $\boldsymbol{\theta}$ by optimizing the following large-scale non-convex optimization problem, which does not admit a closed-form solution:

$$\arg\min_{\boldsymbol{\theta}, f} c(\boldsymbol{\theta}, f; \mathcal{P}, \mathcal{Q}). \tag{11}$$

We propose an approach based on multi-start local (iterative) optimization that proves very effective in practice. We run local optimization from one or more initializations (also called starts), and the solution yielding the lowest overall cost is returned. To reduce computational cost, we propose a mechanism to terminate repeated search of the same search region from different initializations. There is a trade-off in having a larger number of initializations, which increase computational cost but may also increase the chances of finding the global minimum. This trade-off is explored in the experimental section of the chapter.

We now describe how the initialization set is generated and then describe the multi-start optimization algorithm.

3.2.2 Generating the Initialization Set

We define an initialization set $\mathcal{I}$ as $S \geq 1$ pairs: $\mathcal{I} = \{(f_1, \boldsymbol{\theta}_1), \ldots (f_S, \boldsymbol{\theta}_S)\}$, with each pair being an initial focal length and a corresponding initial deformation. We generate $\mathcal{I}$ by exploiting the fact that given an initial focal length, we can initialize deformation reasonably well using an existing closed-form SfT method. We therefore first generate a set of initial focal lengths, and then we pass each of these, together with the other camera intrinsics, the template, and the point correspondences, to a closed-form SfT method, to generate the initial deformations.

Focal Length Generation

We compare two approaches to generate initial focal lengths. The first approach generates one focal length, estimated analytically from the set of point correspondences [3]. This method works best with relatively dense correspondences and smooth, well-textured surfaces. The second approach, which does not depend on the correspondences, works by focal length sampling.

We sample focal lengths using the opening angle representation, which is invariant to image resolution. The focal length f and lens opening angle ψ are related as follows:

$$\tan\left(\frac{\psi}{2}\right) = \frac{s}{2f}, \quad s \stackrel{\text{def}}{=} \max(w, h), \tag{12}$$

where w and h denote the image width and height in pixels, respectively. In real-world SfT applications, lens opening angles are limited by two factors: (i) the physical limits of camera hardware and (ii) theoretical limits and well-posedness of our problem. Concerning (i), the distribution of opening angles of real cameras has been studied previously [45]. The distribution is mono-modal with a mode of approximately 50° and a maximum of approximately 100°, equivalent to a short focal length of $f \approx \frac{1}{2}\max(h, w)$ px. This sets a focal length lower bound in practice. In contrast, *(ii)* sets a focal length upper bound in practice for the following reason. A smaller opening angle (longer focal length) reduces the field-of-view, which in turn causes the viewing rays to become more parallel. When the viewing rays are almost parallel (known as quasi-affine projection), it can be difficult to stably estimate focal length with noise, which is a known result from camera calibration with rigid objects. Consequently, fSfT will not be solvable in real-world cases if the opening angle is very small. As such, we restrict the range of opening angle samples to $20° \leq \psi \leq 100°$. We note that this range is more than sufficient to cover all public datasets that have been used to test previous SfT methods. We found that in practice, we do not need to densely sample focal lengths and good results can be achieved with as few as three samples (20°, 50°, and 80°), corresponding to a narrow, average, and wide field-of-view.

Deformation Generation

Given an initial focal length, there are several closed-form SfT methods that could be used to initialize deformation. We compare two of these. The first is the so-called *maximum depth heuristic* method, referred to as MDH [37, 42]. This makes a convex relaxation of the isometric constraints, leading to a second-order cone programming (SOCP) problem that can be solved with a depth maximization heuristic [37] or with the interior point method [42]. In this chapter, we use the interior-point method implemented in SeDuMi [48]. The second approach uses a perspective-n-point (PnP) method, referred to as PnP, which gives the best-fitting rigid pose. We also compare using both MDH and PnP (generating two initial deformations for each initial focal length). This generates twice as many initializations; however, it is handled efficiently by detecting repeated search during optimization, described in the following section. We find that this works better than using either MDH or PnP alone, and we give implementation details of MDH and PnP in Sect. 4 of the appendix.

3.2.3 Optimization Process and Pseudocode

Our optimization process is summarized in pseudocode in Algorithm 1. Each initialization from the initialization set is processed (either in parallel or sequentially), and local optimization is performed in two steps (lines 7 and 8). At line 7, deformation is optimized with focal length fixed, and at line 8, they are both optimized jointly. These two steps are used to improve convergence especially when the initial focal length is far from the true solution. We implement local optimization with Gauss–Newton with backtracking line search until some termination criteria are satisfied, denoted by T_1 and T_2, respectively. When all initializations have been processed, the solution with lowest cost is taken, and a final refinement is performed with local optimization using termination criteria T_3.

Algorithm 1 fSfT optimization

Require:
$\{(f_1, \boldsymbol{\theta}_1), \dots (f_S, \boldsymbol{\theta}_S)\}$ ▷ initialization set
$c(\boldsymbol{\theta}, f) : \Theta \times \mathbb{R}^+ \to \mathbb{R}^+$ ▷ cost function
1: **function** FSFT_OPTIMIZE($\{(f_1, \boldsymbol{\theta}_1), \dots (f_S, \boldsymbol{\theta}_S)\}, c$)
2: $c^* \leftarrow \infty$ ▷ lowest cost found so far
3: $\mathcal{H} \leftarrow \emptyset$ ▷ search history
4: $f^* \leftarrow 0$, $\boldsymbol{\theta}^* \leftarrow \mathbf{0}$ ▷ best solution with cost c^*
5: **for** $s \in [1, S]$ **do**
6: initialize estimates: $\hat{f} \leftarrow f_s, \hat{\boldsymbol{\theta}} \leftarrow \boldsymbol{\theta}_s$
7: locally optimize c w.r.t. $\hat{\boldsymbol{\theta}}$ until stopping criteria T_1 satisfied.
8: locally optimize c w.r.t. $\hat{\boldsymbol{\theta}}$ and $\hat{f}$ until stopping criteria T_2 satisfied.
9: update history: $\mathcal{H} \leftarrow \mathcal{H} \cup \{(\hat{f}, \hat{\boldsymbol{\theta}})\}$
10: **if** $c(\hat{f}, \hat{\theta}) < C^*$ **then**
11: $(f^*, \boldsymbol{\theta}^*) \leftarrow (\hat{f}, \hat{\boldsymbol{\theta}})$
12: $c^* \leftarrow c(f^*, \boldsymbol{\theta}^*)$
13: Final refinement: locally optimize c initialized with $(f^*, \boldsymbol{\theta}^*)$ until stopping criteria T_3 satisfied.
14: **return** f^* and $\boldsymbol{\theta}^*$

To prevent unnecessary repeated search from different initializations, we maintain a search history $\mathcal{H}$ that holds all the solutions that have been found from a previous initialization (line 9). During the local optimization stages (lines 7, 8, and 13), we continually measure the distance of the current estimate $\hat{f}$ and $\hat{\boldsymbol{\theta}}$ to the closest member of $\mathcal{H}$ using a distance function $d_{\mathcal{H}}((\hat{\boldsymbol{\theta}}, \hat{f}), \mathcal{H})$. We terminate local optimization early if $d_{\mathcal{H}}((\hat{\boldsymbol{\theta}}, \hat{f}), \mathcal{H}) \leq \tau_{\mathcal{H}}$, where $\tau_{\mathcal{H}}$ is a threshold. The distance function is designed to tell us when the current estimate is likely to converge on a solution that already exists in $\mathcal{H}$ (and therefore when we should terminate local optimization). We measure distance in terms of surface normal dissimilarity as follows:

$$d_{\mathcal{H}}((\boldsymbol{\theta}, f), \mathcal{H}) = \min_{(\boldsymbol{\theta}', f') \in \mathcal{H}} \max_{t \in [1,T]} \text{abs}\left(\measuredangle\left(\boldsymbol{n}_t(\boldsymbol{\theta}), \boldsymbol{n}_t(\boldsymbol{\theta}')\right)\right), \tag{13}$$

where $\boldsymbol{n}_t(\boldsymbol{\theta})$ is the surface normal for triangle t generated by $\boldsymbol{\theta}$, and $\measuredangle(\boldsymbol{a}, \boldsymbol{b})$ is the angle in degrees between vectors $\boldsymbol{a}$ and $\boldsymbol{b}$.

The following conditions are used in the termination criteria:

- S1 **Maximum iterations**: The number of iterations τ_{step} has been performed.
- S2 **Small parameter update**: The relative change of all unknowns is below a threshold τ_{Δ}.
- S3 **Small cost update**: The relative change of c is below a threshold τ_c.
- S4 **Out-of-bounds focal length**: $\hat{f}$ is out of bounds: $\hat{f} \leq f_{min}$ or $\hat{f} \geq f_{max}$.
- S5 **Repeated search**: The current solution is similar to one already in the search history: $d_{\mathcal{H}}((\hat{\boldsymbol{\theta}}, \hat{f}), \mathcal{H}) \leq \tau_{\mathcal{H}}$.

S1–S3 are standard in local optimization. S4 is used to terminate early if optimization is converging on a focal length solution that is clearly wrong. Normally, this happens either when the problem is degenerate or when optimization has been very poorly initialized.

The termination criteria T_1, T_2, and T_3 in Algorithm 1 are instantiated by defining thresholds τ_{step}, τ_{Δ}, τ_c, $f_{\min}$, $f_{\max}$, and $\tau_{\mathcal{H}}$. The same values are used in all experiments and are given in Table 1 of the Appendix. We use $f_{\min} = 0.1w$ and $f_{\max} = 1000w$ in all cases, where w is the image width. We highlight why the bounds are different to the focal length bounds defined in Sect. 3.2.2. Those bounds concerned the sampling range for initializing focal length, with opening angles between 20° and 100°. However, there could be cases where the true focal length lies out of these bounds. For that reason, the range of permissible focal lengths during optimization is larger than the range considered for initialization. The range of $0.1w$ to $1000w$ is arbitrary, and it is probably overly broad in practice.

4 Experimental Results

4.1 Datasets

We evaluate our method on 12 public datasets of quasi-isometrically deforming objects from the existing SfT literature (Fig. 4), with a total of 310 test images. These datasets represent a range of real-world challenges, in particular strong deformation and weak texture. Each dataset has a set of images of a deformable object, a

Fig. 4 The 12 public datasets used for evaluation. One representative image per dataset is shown

template, and point correspondences in each image. We give full dataset details, including the number of images per dataset, focal lengths, and the number of points correspondences in Sect. 5 of the Appendix.

The first four datasets ("Spider-man" [9], "Kinect paper" [53], "Van Gogh paper" [43], and "Hulk" [7]) are of smoothly deforming paper sheets that are relatively well textured. The Spider-man dataset has images taken at 9 different focal lengths with opening angles from 24.8° to 65.3°. The other datasets have fixed focal lengths with opening angles 62.4°, 44.5°, and 66.1°, respectively. The next four datasets ("Cap" [2], "Bedsheet" [41], "Kinect t-shirt" [53], and "Handbag" [18]) are of deforming objects made of cloth. These datasets have fixed focal lengths with opening angles of 53.3°, 44.5°, 62.4°, and 50.9°, respectively. The next two datasets ("Floral paper" [18] and "Fortune teller" [18]) are of creased paper objects with sparse texture, making them especially difficult objects. The next dataset ("Bending cardboard" [44]) is of a smoothly deforming cardboard sheet with very sparse texture. The final dataset ("Pillow cover" [18]) is of a deforming pillow cover made of fabric with sparse texture. Outlier-fee point correspondences are provided with six of the datasets (Spider-man, Hulk, Handbag, Floral paper, Fortune teller, and Pillow cover). We generated point correspondences for the other datasets ourselves. The images from the Kinect paper, Van Gogh paper, Bedsheet, Kinect t-shirt, and Bending cardboard are from video clips, so point correspondences were made by tracking keypoints over time. We used KLT feature tracking [51], which worked well in practice because the objects deform relatively smoothly with limited motion blur. Forward–backward consistency checking was used to detect and remove outliers tracks. Point correspondences for the cap dataset were computed by hand using an interactive graphical user interface.

Several of the datasets have images where the object is flat and facing the camera (the Kindet paper, Van Gogh paper, Bedsheet, Kinect t-shirt, and Bending cardboard datasets). These cases are unsolvable because of the ambiguity between focal length and surface depth. We therefore exclude an image from the evaluation if all surface normals approximately align with the optical axis (we use a threshold of 5°).

4.2 Evaluation Metrics

We evaluate solution accuracy for each image with two metrics. The first is focal length percentage error (FLPE) and the second is shape error (SE). FLPE is defined as follows:

$$\mathrm{FLPE}(\hat{f}, f) \stackrel{\text{def}}{=} 100 \times \frac{\left|\hat{f} - f^{gt}\right|}{f^{gt}} \tag{14}$$

where $\hat{f}$ is the estimate focal length and f^{gt} is the ground truth focal length (all datasets provide ground truth focal lengths). SE is computed for all datasets with ground truth (Spider-man, Kinect paper, Hulk, Cap, Kinect t-shirt, Handbag, Floral paper, Fortune teller, and Bedsheet) as follows. For each image and each point correspondence, we evaluate the Euclidean distance between the reconstructed 3D point in camera coordinates $\hat{\boldsymbol{q}} \in \mathbb{R}^3$ and ground truth $\boldsymbol{q}^{gt} \in \mathbb{R}^3$. The *reconstruction error* (RE) is defined as $\mathrm{RE}\left(\hat{\boldsymbol{q}}, \boldsymbol{q}^{gt}\right) \stackrel{\text{def}}{=} \left\|\hat{\boldsymbol{q}} - \boldsymbol{q}^{gt}\right\|$.

RE has been used extensively for SfT evaluation. However, it has an important limitation for fSfT evaluation. We now explain this, motivating the use of an adapted metric, called the shape error (SE). The isometric prior penalizes stretching and shrinking of the template. This fixes the scale ambiguity that would otherwise exist between f and the template's scale. However, in cases where the perspective effects are weak (i.e., when the viewing rays of the point correspondences were approximately parallel), an ambiguity emerges between f and the template's average depth $\bar{d}$. That is, one can obtain a similar image by reducing f by a scale factor α and increasing $\bar{d}$ by $\frac{1}{\alpha}$. This ambiguity is well-known in the case of rigid objects and was previously identified in fSfT [3]. In terms of evaluation, this highlights a shortcoming of RE: in cases with weak-perspective effects, it may be possible to reconstruct the shape of the template accurately, but not possible to precisely determine $\bar{d}$ and f. A method able to accurately reconstruct shape in these cases would receive a high RE error, which would be unfair.

To handle this, we adapt RE to make it insensitive to a global shift in average depth. First, a least-squares translation t_z is computed along the camera's optical axis to align the reconstructed 3D point correspondences with their ground truths. SE is then computed as follows:

$$\mathrm{SE}\left(\hat{\boldsymbol{q}}, \boldsymbol{q}^{gt}\right) \stackrel{\text{def}}{=} \frac{100}{S} \times \left\|\hat{\boldsymbol{q}} + t_z - \boldsymbol{q}^{gt}\right\|_2. \tag{15}$$

The denominator S is used to make SE independent of the template's size. We set this as the maximum spatial range of the template's rest shape with respect to its 3 spatial coordinates. Consequently, an SE of 1 corresponds to approximately 1% of the template's size. We emphasize that this is to help interpret results, and it is not linked to a reconstruction scale ambiguity.

4.3 Success Rates

We compare performance using *success rates*, which are the proportion of images for which a method returns a solution with an error less than a threshold τ. We use FLPE-success@τ to denote the FLPE-success rate using a threshold τ. Similarly, we use SE-success@τ to denote the SE-success rate. We use a few different thresholds to assess how often very accurate results are achieved (smaller τ) and how often results in the right "ballpark" are achieved (larger τ). Success rate was selected because it is a robust statistic, required to handle the fact that in some instances fSfT can be weakly posed, leading to extreme FLPE and SE values.

4.4 FLPE and SE Results

We evaluate our approach with three different policies for creating the initialization set. There is a trade-off between using a larger initialization set (increasing computational cost) and a smaller initialization set (reducing computational cost but potentially reducing the chance of finding the global optimum). In this section, we test three initialization policies that are specified by a set of lens opening angles, defined as Ψ_{init}, and a set of closed-form SfT methods, defined as $\mathcal{M}$:

- Policy 1: $\Psi_{init} = \{\psi^{An}\}$, $\mathcal{M} = \{\text{MDH, PnP}\}$.
- Policy 2: $\Psi_{init} = \{20, 50, 80\}$, $\mathcal{M} = \{\text{MDH, PnP}\}$.
- Policy 3: $\Psi_{init} = \{20, 30, 40, 50, 60, 70, 80\}$, $\mathcal{M} = \{\text{MDH, PnP}\}$.

We use ψ^{An} to denote the opening angle estimated by the analytical method. For each focal length initialization, we generate two deformation initializations using MRD and PnP. The numbers of initializations S for policies 1, 2, and 3 are therefore 2, 6, and 14, respectively. We compare results against the analytical method to solve focal length, combined with the MDH method to compute deformation. This combination is denoted as **FAn + MDH**.

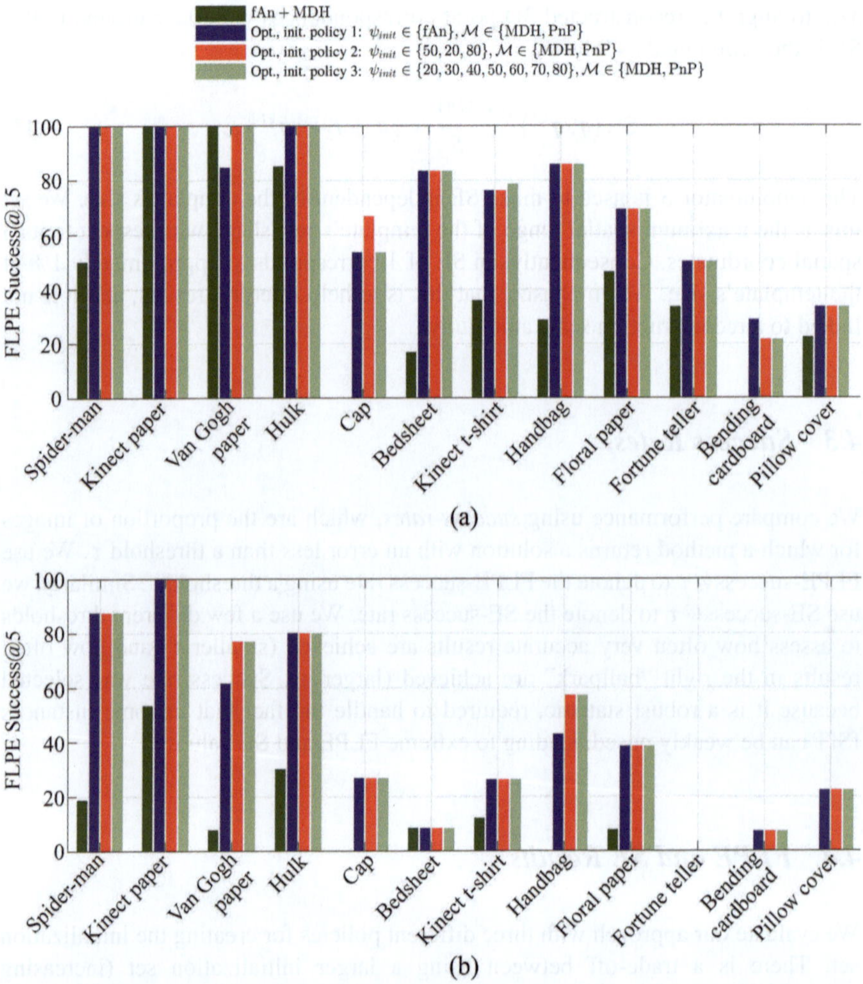

Fig. 5 Focal length percentage error (FLPE) results for the analytical method (denoted as "fAn + MDH") and optimization-based method (denoted as "Opt.") using different initialization policies. The initialization policies are defined in terms of the set ψ_{init} of initial focal lengths and the set of SfT methods $\mathcal{M}$ used to initialize deformation. (**a**) Shows FLPE-success rates at 15% and (**b**) shows FLPE-success rates at 5%

We consider FLPE below 15% to be a good result for fSfT and FLPE below 5% to be an exceptional result. Thus, we evaluate both FLPE-success@15 and FLPE-success@5, shown in Fig. 5a, b, respectively. Similarly, Fig. 6a, b shows SE-success@5% and SE-success@2%, respectively.

We first consider FLPE-success@15 in Fig. 5a. We observe the following points:

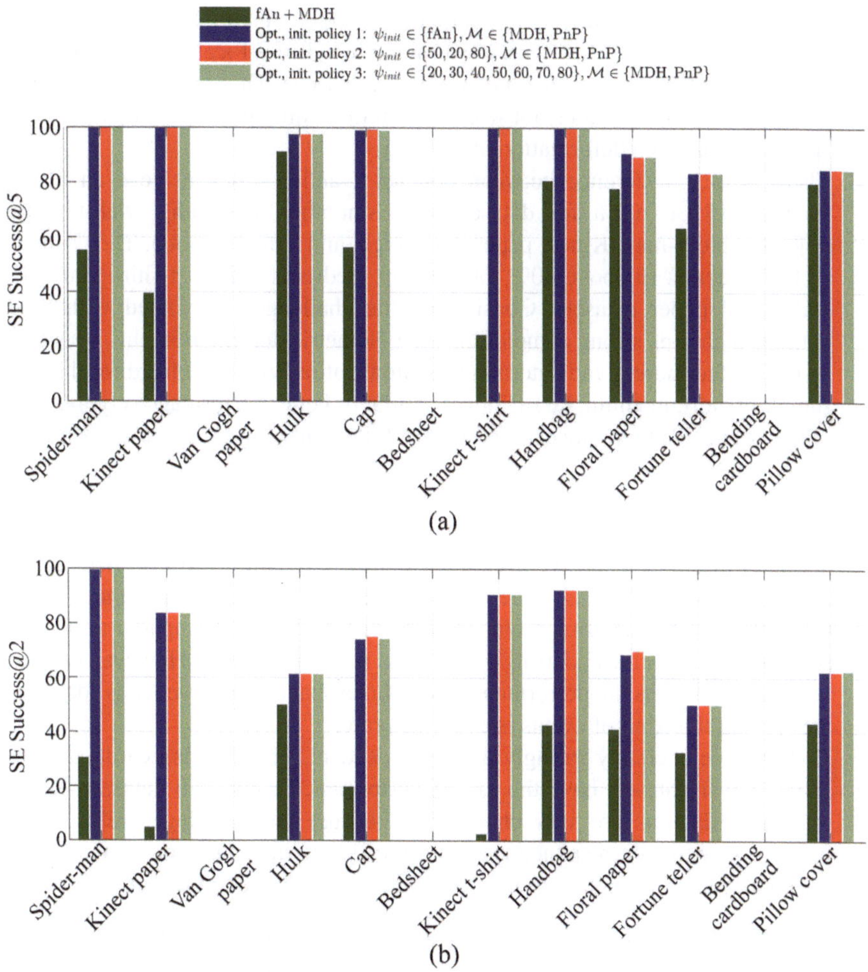

Fig. 6 (**a** and **b**) Shows the shape error (SE) of the method (denoted as "fAn + MDH") and optimization-based method (denoted as "Opt.") using different initialization policies. The initialization policies are defined in terms of the set ψ_{init} of initial focal lengths and the set of SfT methods $\mathcal{M}$ used to initialize deformation. Van Gogh paper, Bedsheet, and Bending cardboard datasets have no errors because they do not contain ground truth 3D information. For this reason, there are no bars associated with them

1. The performance of **FAn + MDH** is very good for the Kinect and Van Gogh Paper datasets, where FLPE-success@15 is 100.0%. **FAn + MDH** achieves a relatively high FLPE-success@15 of 80.0% for the Hulk dataset. Recall that these datasets are smoothly deforming paper sheets with dense texture. These results indicate that the analytical method can estimate the focal length well in these cases.
2. For the other datasets (Spider-man, Cap, Bedsheet Kinect t-shirt, Handbag, Floral paper, and Bending cardboard), **FAn + MDH** performs relatively poorly and much worse than optimization-based method (with any initialization policy).

Indeed for the Cap and Bending cardboard datasets, **FAn + MDH** has a FLPE-success@15 of 0.0%: therefore, it was not able to find a focal length within 15% of ground truth in any of the images of those datasets. These results indicate that the analytical method does not work well in more difficult cases when texture is sparse and/or when deformation is complex.

3. There is little difference between policies 2 and 3. They achieve an FLPE-success@15 of 100% for datasets with smoothly deforming, well-textured objects (Spider-man, Kinect paper, Van Gogh, and Hulk datasets). They achieve FLPE-success@15 above 60% for the Cap, Bedsheet, Kinect t-shirt, Handbag, and Floral paper datasets. Considering the challenges associated with these datasets including strong complex and non-isometric deformation, this is a strong result. Furthermore, it indicates that: (i) initialization with three fixed focal length samples (short, medium, far) achieves similar or better performance compared to policy 1 and (ii) there appears to be very little benefit in using more than three focal length samples.
4. For the Cap dataset, policy 2 has a higher success rate than policy 3. This may seem surprising because policy 3 initializes with more starts, including all starts in policy 2, so we may think policy 3 should always do better. This is not necessarily the case. The reason is because there exists an image in the Cap dataset with a spurious solution that has a lower cost compared to the true solution. This was located using policy 3 but not with policy 2. However, because in all other datasets the performances of policies 2 and 3 are practically identical, we see that this kind of events is extremely rare.
5. Performance is clearly strongly dataset dependent. The Bending cardboard and Pillow cover datasets have the lowest performance among all datasets. Recall that these datasets are very challenging because the Bending cardboard has extremely sparse correspondences, and the Pillow cover has many views that are approximately fronto-parallel (making fSfT poorly conditioned).

We now consider FLPE-success@5 in Fig. 5b. We observe the following:

6. Because of the much more stringent success threshold of 5%, we observe lower success rates for most datasets. Nevertheless, 100% success rate is achieved by the optimization-based method for the Kinect paper dataset with all initialization policies. Success rates above 78% are achieved for the Spider-man, Van Gogh paper, and Hulk datasets with the optimization-based method and all initialization policies. This shows that we can solve fSfT with the optimization-based method and achieve very high accuracy (FLPE below 5%) for strongly isometric and well-textured objects.
7. For less isometric and/or weakly textured objects (those other than Spider-man, Kinect paper, Van Gogh paper, and Hulk datasets), it is very challenging to solve fSfT consistently with high accuracy and FLPE below 5%.
8. Unlike FLPE-success@15, **FAn + MDH** achieves significantly lower success rates for FLPE-success@5 with the Kinect paper, Van Gogh, and Hulk datasets compared to the optimization-based method. This indicates that the analytical

method can achieve focal lengths in the right ballpark (<15% error) for isometric well-textured objects, but it is not as precise as the optimization-based method.

We now consider shape error (SE) shown in Fig. 6. Recall that Van Gogh, Bedsheet, and Bending cardboard datasets do not have ground truth 3D information so SE cannot be measured. We observe the following points:

9. Very similar SE results are achieved for the different initialization policies for each dataset. This agrees with the FLPE results.
10. SE-success@5 is above 80% for all datasets with the optimization-based method with all initialization policies. Recall that an SE of 5 occurs when the Euclidean error at each reconstructed point is within 5% relative to the size of the object template. Thus, SE-success@5 above 80% is a strong result.
11. Unlike the FLPE results, the simpler datasets (Spider-man, Kinect paper, and Hulk) do not have systematically better SE results compared to other datasets. Indeed, the Pillow cover dataset, which had the second lowest FLPE-success rate among all datasets, has very similar SE-success@2 as the Hulk dataset. This highlights the intrinsic difficulty of the Pillow cover dataset. It has little variation in depth, leading to weak-perspective effects. This causes an ambiguity between the distance of the object to the camera and focal length, explaining why its shape can be estimated well but the focal length cannot.
12. The shape error of **FAn + MDH** is similar to the optimization-based method for the Hulk dataset, but it is generally significantly worse for the other datasets. Recall that the Hulk dataset has relatively strong-perspective effects thanks to the short focal length and deformation is smooth with well-distributed point correspondences (Fig. 7).

4.5 Results Visualizations

We visualize results of the optimization-based method (policy 2) in Figs. 8, 9, and 10. The figures are laid out in the same way, with four representative images per dataset. Below the images are the corresponding deformation solutions rendered from the camera's viewpoint and shaded to show the shapes. We give the corresponding FLPE with each image below each render. For all datasets with ground truth, point correspondences are colored-mapped using their shape error. For datasets without ground truth, point correspondences are colored in green. We can see the method has been able to estimate deformations very well, especially considering the complex deformations exhibited by the handbag, floral paper, fortune teller, and cap datasets. Note that the floral paper and fortune teller datasets have strongly creased objects with relatively few point correspondences. We see our method is able to recover the general shape well despite these challenges. In some datasets, a relatively large FLPE was obtained (mainly the cardboard and pillow cover datasets), while the shape appears to be reconstructed well. These results

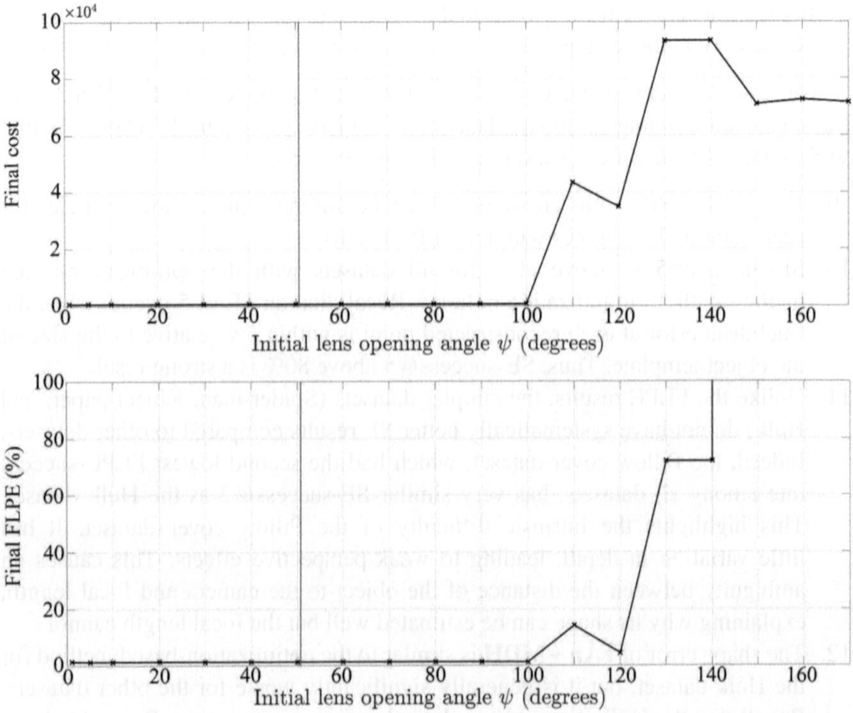

Fig. 7 Graphical illustration of the cost function's convergence basin with respect to initial focal length (expressed in opening angles). 20 different initial focal lengths are tested using Algorithm 1, ranging from 1° (extreme tele-photo) to 170° (extreme wide angle). The top graph shows final cost, and the bottom graph shows final FLPE against initial focal length, respectively. Ground truth focal length is shown as a vertical line

suggest fSfT is weakly posed in those cases, with an ambiguity between focal length and surface distance.

4.6 Convergence Basin

As only a few initializations are required with widely spread focal lengths to achieve good results, this suggests that the cost function's convergence basin is relatively wide. We provide a graphical illustration of the convergence basin in Fig. 7 using the first image from the Floral paper dataset as a typical example. These graphs are generated by sampling 20 focal lengths with opening angles ranging from 1° (extreme tele-photo) to 170° (extreme wide angle), including the ground truth focal length. For each sample, Algorithm 1 is run using two SfT methods (MDH and PnP), and the cost of the final solution returned by the algorithm is recorded. Figure 7 (top)

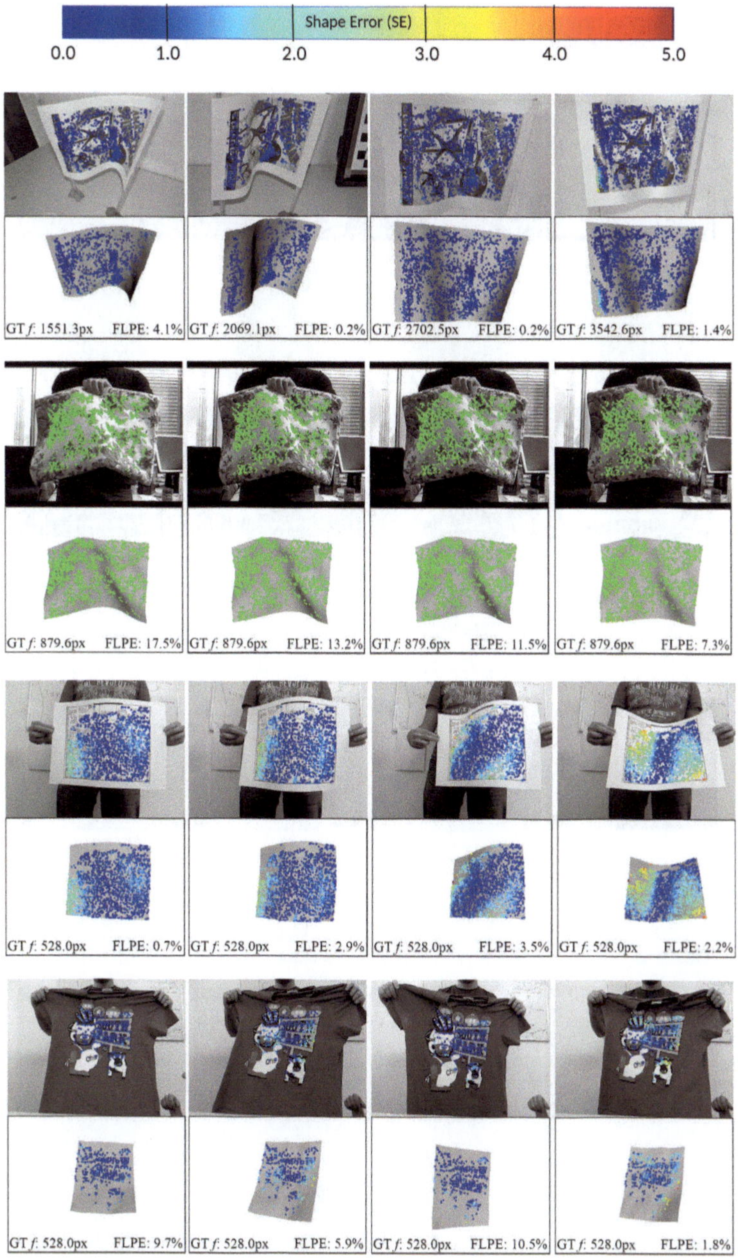

Fig. 8 fSfT results with our method. Four representative images from the Spider-man, Bedsheet, Kinect paper, and Kinect t-shirt datasets are shown (top to bottom). The estimated deformations are shown below each image as shaded renders. Point correspondences are visualized and color-mapped according to their SE. The Bedsheet dataset does not have ground truth 3D, so point correspondences are shown in green. Focal length information and FLPE are given below each render

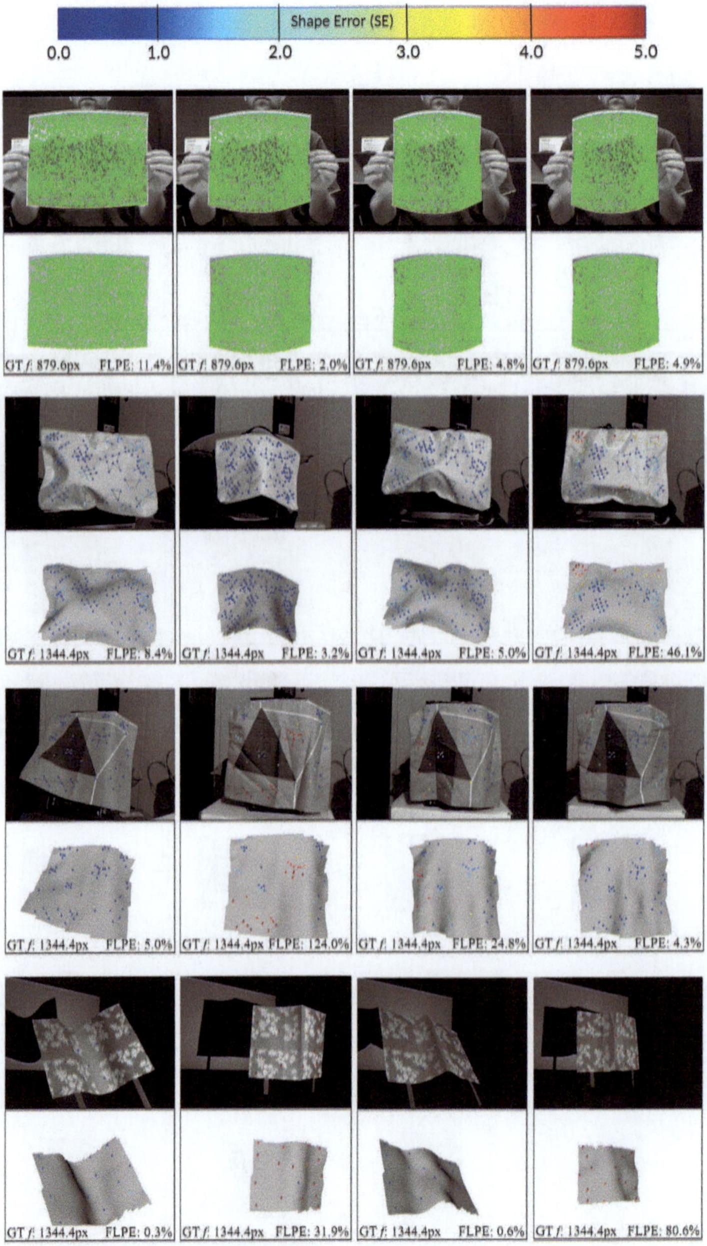

Fig. 9 fSfT results with our method. Four representative images from the Van Gogh paper, Handbag, Pillow cover, and floral paper datasets are shown (top to bottom). The estimated deformations are shown below each image as shaded renders. Point correspondences are visualized and color-mapped according to their SE. The Van Gogh paper dataset does not have ground truth 3D, so point correspondences are shown in green. Focal length information and FLPE are given below each render

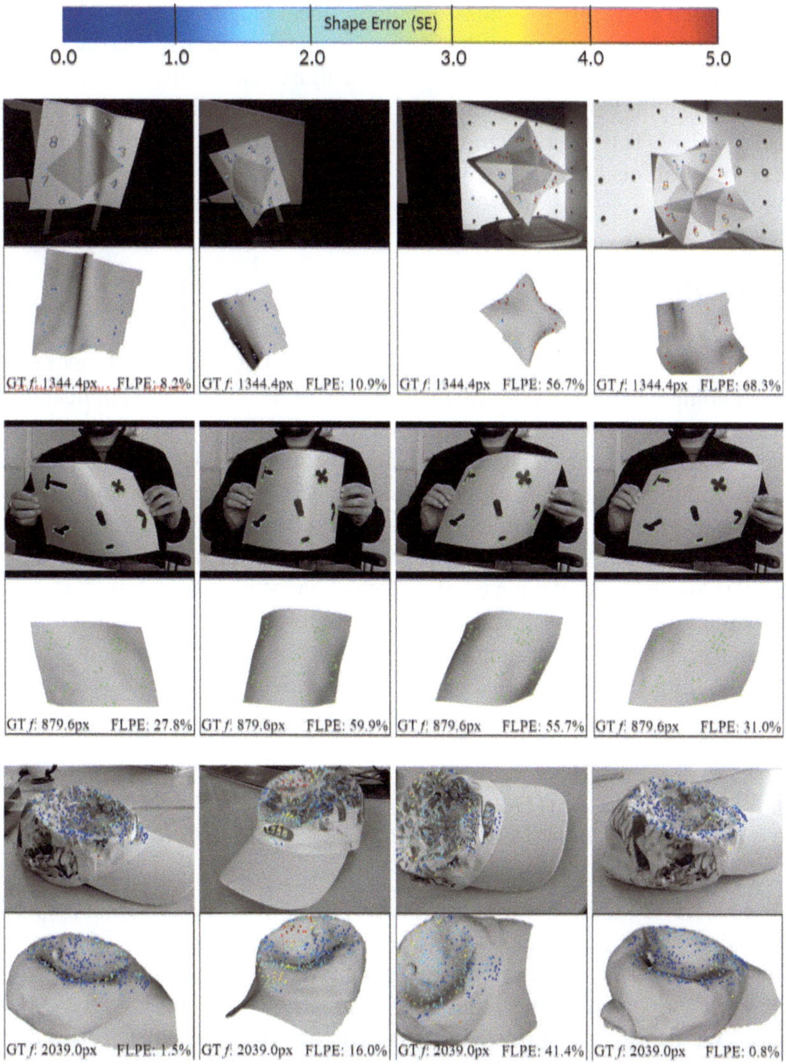

Fig. 10 fSfT results with our method. Four representative images from the Fortune teller, Bending cardboard, and Cap datasets (top to bottom). The estimated deformations are shown below each image as shaded renders. Focal length information and FLPE are given below each render

shows the final costs plotted against the initial focal lengths (expressed as opening angles). Figure 7 (bottom) shows the FLPE of the final solution plotted against initial focal length. The ground truth focal length is illustrated by the vertical lines. For initial focal lengths between 1° and 100°, the same FLPE and final cost are achieved (FLPE of 1.2%). This clearly demonstrates a relatively wide convergence basin with respect to initial focal length.

4.7 Results Summary

The results show that the optimization-based method generally achieves far better accuracy compared to the analytical method. Therefore, in practical applications, the analytical method should be considered as a way to initialize the optimization-based method (as done in policy 1) and not as a competitive approach. Initialization with the analytical method appears to achieve similar accuracy compared to initialization with focal length sampling (policies 2 and 3). Furthermore, there appears to be no benefit in initializing with more than three focal length samples (policy 2 versus policy 3). We have implemented the method in Matlab a standard x64 Linux workstation computer, and it runs in approximately 5 s using initialization policy 2 (6 initializations). We confidently believe it can be run in real time with a C++ and CUDA implementation based on [12]. A more detailed analysis of computation time with different initialization policies is provided in Sect. 7 of the appendix.

4.8 Additional Initialization Sensitivity Experiments

We conducted an additional experiment to further investigate the impact of the initialization policy on computation time and accuracy. The public datasets were limiting for this because the range of opening angles was not very large (24.8° to 65.3°). Consequently, one focal length sample at 50.0° worked very well in practice. We therefore augmented the datasets by adding simulated digital zoom variability. We also added synthetic point correspondence noise. Because of space restrictions, we give further details for the augmented datasets and results in Sect. 6 of the appendix, and we summarize the findings here:

13. In the original datasets (without simulated digital zoom or point correspondence noise), initialization with the analytical solution had very similar performance as initialization with one focal length sample at 50.0°. However, when zoom augmentation was added, initialization with the analytical solution outperformed one focal length sample at 50.0°.
14. Initializing with three opening angles of 20.0°, 50.0°, and 80.0° gives slightly better performance compared to initializing with analytical solution.

15. There is a clear benefit in using both PnP and MDH for initializing deformation, compared to using just MDH. Because the optimization algorithm had an early termination mechanism to avoid repeated search, the computational cost of using both PnP and MDH is only approximately 50% greater than just using MDH.
16. There is little benefit in using more than three focal length samples.

We therefore recommend initializing with three opening angles at 20.0°, 50.0°, and 80.0° as the default and using both PnP and MDH for initializing deformation for each opening angle. We refer the reader to Sect. 6 of the appendix for the experiment details and quantitative performance statistics.

4.9 Isometric Weight Sensitivity

We end the experiments section by investigating the sensitivity of the isometric weight λ_{iso} on solution accuracy and showing the positive effects our normalization techniques have had in restricting the range of good isometric weights. We test 10 different isometric weights ranging from $\lambda_{\text{iso}} = 1000 \times \lambda'_{\text{iso}}$ to $\lambda_{\text{iso}} = \frac{1}{100} \times \lambda'_{\text{iso}}$, given in the legend in Fig. 11. For each isometric weight, we optimize the cost function using Algorithm 1, initialized using initialization policy 3 (3 opening angles at 20.0°, 50.0°, and 80.0°, and using both PnP and MDH for initializing deformation). The results are shown in Fig. 11. We see that λ_{iso} strongly influences performance, where excessively large or small weights compared to λ'_{iso} lead to poor performance. This is seen for all dataset groups and performance metrics. Performance is generally uni-modal in λ_{iso}, and peaking at, or very close to λ'_{iso}. Using a fixed weight in the range $\frac{1}{2}\lambda'_{\text{iso}} \leq \lambda_{\text{iso}} \leq 2\lambda'_{\text{iso}}$ leads to similar performance where $\lambda_{\text{iso}} = \lambda'_{\text{iso}}$ is generally the best. The results also show the importance of normalization. If, for example, we changed the template's size by a factor of 10, then without scale normalization, the influence of the isometric cost (Eq. 3.1.2) would be 10 times greater, and λ'_{iso} would no longer be a good weight: results would be equivalent to using a weight of $10\lambda'_{\text{iso}}$, with much worse results. Similarly, if we were to not normalize the number of point correspondences N, a change in N by a factor of 10 would also lead to significantly worse performance equivalent to $\lambda_{\text{iso}} = \lambda'_{\text{iso}}/10$. These results strongly indicate our normalization strategies are effective, because the same isometric weight can be used in all datasets with near-optimal performance.

5 Conclusion

fSfT is a challenging problem that must be solved in real-world SfT applications when the camera's focal length (a key intrinsic) is unknown. We have modeled and solved fSfT with a large-scale non-convex cost function that is optimized

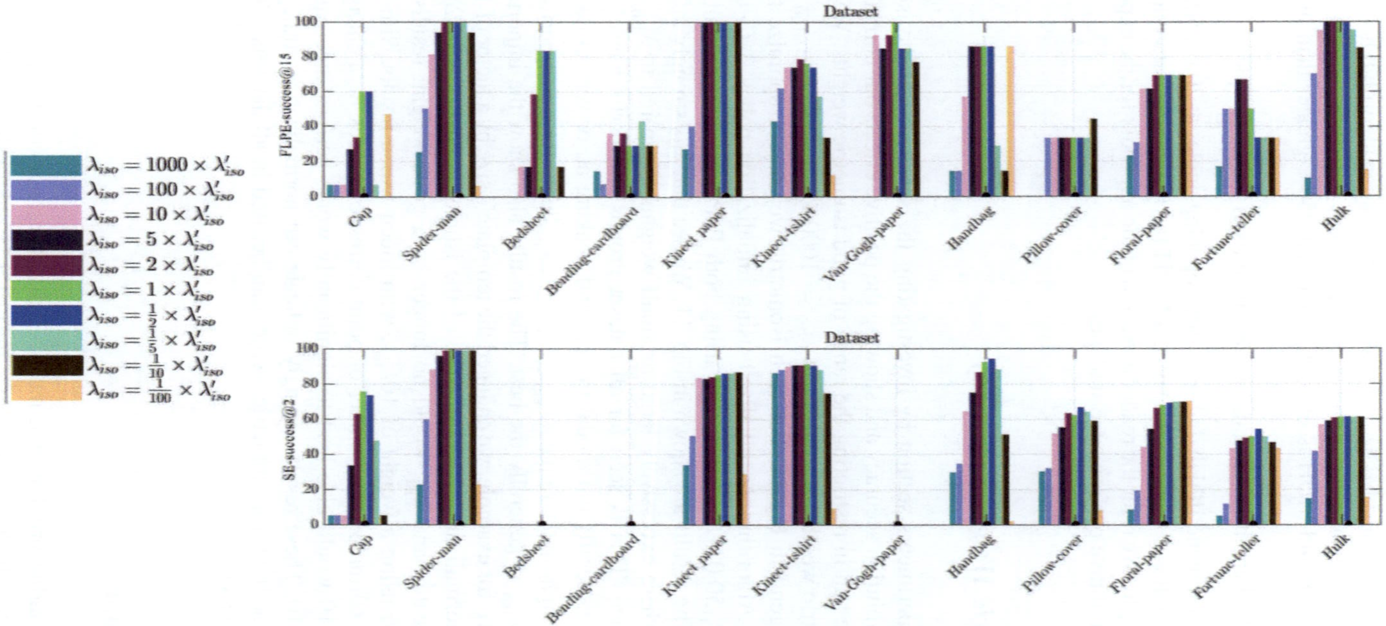

Fig. 11 FLPE performance (top) and SE performance (bottom) with different isometric weights. Note that Bedsheet, Bending cardboard, and Van Gogh paper do not have ground truth 3D, so SE for those datasets cannot be measured

with multi-start Gauss–Newton with a mechanism to avoid repeated search. The method has received a relatively large evaluation comprising 12 public datasets with various challenges that include sparse texture and creased surfaces. The method is considerably more accurate compared to the analytical solution [3] in all datasets. It uses a carefully designed cost function with normalization techniques that allow the same deformation weights to be used for all datasets, making our solution practical for real-world use without the need to tune cost weights at run time. In future work, we aim to combine the method with deep-learning-based dense image registration and to extend the method to multiple views sharing a common focal length. We also aim to study the practical and theoretical limits of solving SfT with more unknown intrinsics.

Matching quality is an important factor that affects results, and a deeper analysis would be useful. There are three main aspects to match quality: match density, match noise, and incorrect matches (also called "outliers" or "mismatches"). By including objects with different amounts of texture, match density is varied in our experiments. As shown in Fig. 13, the poorly textured objects (Floral paper, Fortune teller, Bending cardboard, and Pillow cover) tend to have higher reconstruction error than well-textured objects (with denser matches). Fortune teller represents a "breakpoint" because of the combination of very sparse matches and highly discontinuous deformation (folds). The matches do not provide enough motion information for consistent and accurate focal length estimation. This is shown in Fig. 5a, where focal length estimates with 15% error were attained in 50% of the Fortune teller images. Concerning mismatches, as with most prior works in SfT, we assume that the majority of mismatches have been detected and removed by a dedicated "outlier rejection" method, and then the remaining matches are passed to our fSfT method. In particular, [38] and its extension to arbitrary template meshes [36] have shown excellent robustness in a tracking-by-detection setting. A dedicated evaluation of the robustness of the complete pipeline, using different outlier rejection methods, would be interesting to perform for future work. We also aim to conduct a more detailed analysis of the influence of point noise on reconstruction accuracy.

Appendix

1 Overview

This appendix is organized into 7 sections. Section 2 describes the implementation of the isometric cost defined in Sect. 3.1.2 with a triangulated mesh. Section 3 gives the default values of the termination condition thresholds used in Algorithm 1 and in all experiments. Section 4 gives the implementation details of the MDH and PnP SfT methods. Section 5 provides additional numerical details of each dataset.

Section 6 provides the details for the initialization sensitivity experiment described in Sect. 4.8. Section 7 provides additional computational cost analysis.

2 Discrete Quasi-isometric Cost Implementation

2.1 Triangle Geometry and Embedding Functions

A constant strain triangle (CST) is a triangular element whose stress and strain fields are constant in the triangle's domain. Each triangle is associated with a flat 2D triangular domain $\Omega_t \subset \mathbb{R}^2$, where $t \in [1, T]$ is the triangle index. We define Ω_t by three 2D vertices denoted as $u_t^1 \in \mathbb{R}^2$, $u_t^2 \in \mathbb{R}^2$, and $u_t^3 \in \mathbb{R}^2$. We define as $\boldsymbol{y}_t^i \in \mathbb{R}^3$ and $\boldsymbol{x}_t^i \in \mathbb{R}^3$ the position of the ith vertex of triangle t in object and camera coordinates, respectively. We define as $\zeta_t : \Omega_t \to \mathbb{R}^3$ the embedding of triangle t into object coordinates. We define as $\phi_t : \Omega_t \to \mathbb{R}^3$ the embedding of triangle t into camera coordinates. Recall that ζ_t is known from the template and ϕ_t to be estimated by fSfT.

2.2 Cost

The isometric cost c_{iso} is constructed by the following discrete approximation of E_{strain}:

$$c_{\text{iso}} = \sum_{t=1}^{T} a_t \left\| G\left(J(\zeta_t)\right) - G\left(J(\phi_t)\right) \right\|_F^2 \approx E_{strain}, \tag{16}$$

where $G(X) \stackrel{\text{def}}{=} X^\top X$ is the Gramian operator, a_t is the known surface area of the tth triangle, and $J(f)$ denotes the Jacobian of a function f. Taken strictly, Eq. (16) equates the first fundamental form of the two triangle embeddings.

We position the vertices $\boldsymbol{u}_t^{i \in [1,3]}$ such that the triangle in object coordinates is isometric with Ω_t as follows:

$$\boldsymbol{u}_t^i = \begin{bmatrix} 1 & 0 & 0 \\ 0 & 1 & 0 \end{bmatrix} \boldsymbol{R}_t \, \boldsymbol{y}_t^i, \tag{17}$$

where $\boldsymbol{R}_t$ is any 3D rotation that aligns the triangle's normal in object coordinates to the z axis. As a consequence, $G(J(\zeta_t)) = \boldsymbol{I}_2$. Returning to Eq. (16), we are left with expressing $J(\phi_t)$. Because ϕ_t is linear within Ω_t from the definition of a CST, as a

Table 1 Default termination values used in Algorithm 1

	τ_{step}	τ_Δ	τ_c	$\tau_\mathcal{H}$
T_1	10	1e − 5	1e − 5	20
T_2	20	1e − 5	1e − 5	20
T_3	100	1e − 5	1e − 5	20

local affine embedding $J(\phi_t) \in \mathbb{R}^{3\times 2}$ is constant and it is found from the following linear system of 6 equations:

$$J(\phi_t)\left(\boldsymbol{u}_t^i - \frac{1}{3}\left(\boldsymbol{u}_t^1 + \boldsymbol{u}_t^2 + \boldsymbol{u}_t^3\right)\right) = \left(\boldsymbol{x}_t^i - \frac{1}{3}\left(\boldsymbol{x}_t^1 + \boldsymbol{x}_t^2 + \boldsymbol{x}_t^3\right)\right), \quad \forall i \in [1, 3]. \tag{18}$$

The solution to $J(\phi_t)$ in Eq. (18) is a linear expression in the unknown vertex positions $\boldsymbol{x}_t^1$, $\boldsymbol{x}_t^2$, and $\boldsymbol{x}_t^3$. We recall that from the definition of $\boldsymbol{\theta}$ in Eq. (2), $\boldsymbol{x}_t^1$, $\boldsymbol{x}_t^2$, and $\boldsymbol{x}_t^3$ are contained within $\boldsymbol{\theta}$ ($\boldsymbol{\theta}$ holds the positions of all vertices in camera coordinates). As a consequence, $J(\phi_t)$ is linear in $\boldsymbol{\theta}$, and therefore, Eq. (16) is quartic in $\boldsymbol{\theta}$.

3 Optimization Termination Conditions

4 SfT Implementation Details

4.1 MDH

MDH is considered one of the best closed-form SfT methods. Given an initial opening angle ψ_s with focal length f_s, and the set of N point correspondences $\mathcal{P}$ and $\mathcal{Q}$, we compute $\boldsymbol{\theta}_s$ in two stages as follows. In the first stage, we reconstruct the depths $\mathcal{Z} = \{z_1, z_2, \ldots, z_N\}$ of $\mathcal{Q}$ by solving a convex relaxation of SfT following [42]. Specifically, we maximize the depth of each point such that the Euclidean distance e_{ij} between any two points $(i, j) \in [1, N]^2$ does not exceed their geodesic distance d_{ij}. The geodesic distances are known *a priori* from the template, and they can be computed efficiently using, e.g., Fast Marching [24]. The following SOCP problem is then solved:

$$\begin{aligned} &\max \textstyle\sum_{i=1}^N z_i \quad \text{s.t.} \\ &\forall (i, j) \in \mathcal{N} \quad \left\| \tfrac{z_i}{f_s} \operatorname{stk}(\mathcal{Q}(i), f_s) - \tfrac{z_j}{f_s} \operatorname{stk}(\mathcal{Q}(j), f_s) \right\|_2^2 \le d_{ij}. \end{aligned} \tag{19}$$

The set $\mathcal{N}$ defines pairs of point correspondences, constructed with a K-nearest neighbor (KNN) graph with a default of $\min(N, 15)$ neighbors. We solve Eq. (19) quickly using the interior point method from Mosek [34]. When N is not large

($N \leq 500$), this typically takes between 100 and 500 ms on the benchmark computer. If $N > 500$, we reduce the problem size by randomly sub-sampling 500 correspondences without replacement using furthest point sampling, and we ignore the remaining points. This normally has little effect on reconstruction accuracy.

In the second stage, we compute $\boldsymbol{\theta}_s$ from $\mathcal{Z}$ and f_s. We solve a regularized linear least-squares system that finds a smooth 3D deformation of the template mesh that fits the reconstructed point correspondences in camera coordinates. This problem is as follows:

$$\begin{aligned} \boldsymbol{\theta}_s &= \arg\min_{\boldsymbol{\theta}} \frac{1}{N} \sum_{i=1}^{N} \left\| g(\mathcal{P}(i); \boldsymbol{\theta}) - \frac{z_i}{f} \mathrm{stk}(\mathcal{Q}(i), f_s) \right\|_2^2 + \lambda c_{\mathrm{reg}}(\boldsymbol{\theta}) \quad (a) \\ &= \arg\min_{\boldsymbol{\theta}} \| \boldsymbol{A}_{mdh} \boldsymbol{\theta} - \boldsymbol{b}_{mdh} \|_2^2 \quad (b), \end{aligned} \tag{20}$$

where λ is a regularization weight. Equation (20b) is equivalent to Eq. (20a), where we have rearranged the problem to a standard LLS format with known terms $\boldsymbol{A}_{mdh}$ and $\boldsymbol{b}_{mdh}$. The matrix $\boldsymbol{A}_{mdh}$ does not depend on f_s nor $\mathcal{Z}$. We exploit this by solving Eq. (20) with a factorization of $\boldsymbol{A}_{mdh}$. Importantly, the factorization can be done once and be reused for any f_s or $\mathcal{Z}$. We weight c_{reg} using the normalization technique described in Sect. 3.1.3, and we use the same λ for all problem instances (we use a default of $\lambda = 100$ in all experiments).

The factorization can be solved very quickly when the number of mesh vertices V is small (a few hundred) using sparse Cholesky factorization. However, for larger meshes, it becomes unreasonably expensive. We deal with this by applying dimensionality reduction by eliminating high-frequency deformation components from the problem. We implement this with linear bases using a modal analysis of the design matrix of the regularization cost. This reduces the problem to a smaller dense linear system that is solved efficiently with Cholesky factorization (we use Eigen's LDLT implementation).

4.2 PnP

PnP estimates the rigid pose of the template in camera coordinates from a focal length sample f_s, and the point correspondences $\mathcal{P}$ and $\mathcal{Q}$. Despite not estimating deformation, we find this is a surprisingly effective and fast initialization method for fSfT. When $\mathcal{P}$ is co-planar, we use IPPE [11]; otherwise, we use OpenCV's SolvePnP method that implements the direct linear transform (DLT) initialization and Levenberg–Marquardt refinement.

5 Dataset Descriptions

Additional dataset descriptions are provided in Tables 2, 3, and 4.

Table 2 Cap, Pillow cover, Handbag, and Spider-man dataset statistics

	Cap	Handbag	Pillow cover	Spider-man
Object material	Fabric	Fabric	Fabric	Paper
Template geometry	3D open	3D open	3D open	Flat open
Number of template vertices (V)	4854	1098	1368	2918
Number of template triangles (T)	9502	2063	2587	5000
Video (vid.) or image collection (col.)	Col.	Col.	Col.	Col.
Number of images (M)	15	7	9	79
Image resolution ($w \times h$)	2048×1536	1280×960	1280×960	1728×1152
Correspondences per image (N)	266	150	63	1176 ± 468
Focal length (px)	2039	1344.0	1344.0	$1348.4 \rightarrow 3937.9$
Focal length (% of w)	99.6	105.0	105.0	$78.0 \rightarrow 277.9$
Lens opening angle (°)	53.3	50.9	50.9	$65.3 \rightarrow 24.8$
Has ground truth 3D	Yes	Yes	Yes	Yes

Table 3 Floral paper, Fortune teller, Hulk, and Bending cardboard dataset statistics

	Floral paper	Fortune teller	Hulk	Bending cardboard
Object material	Paper	Paper	Foam	Cardboard
Template geometry	3D open	3D open	Flat open	Flat open
Number of template vertices (V)	1248	936	122	609
Number of template triangles (T)	2342	1747	200	1120
Video (vid.) or image collection (col.)	Col.	Col.	Col.	Vid.
Number of images (M)	13	6	20	18 (87)
Image resolution ($w \times h$)	1280×960	1280×960	4928×3264	720×576
Correspondences per image (N)	18	20	20	52
Focal length (px)	1344.0	$1344.0 \rightarrow 3937.9$	3784.9	879.6
Focal length (% of w)	105.0	$105.0 \rightarrow 277.9$	76.8	122.2
Lens opening angle (°)	50.9	50.9	66.1	44.5
Has ground truth 3D	Yes	Yes	Yes	No

Table 4 Bedsheet, Kinect t-shirt, Kinect paper, and Van Gogh paper dataset statistics

	Bedsheet	Kinect t-shirt	Kinect paper	Van Gogh paper
Object material	Fabric	Fabric	Paper	Paper
Template geometry	Flat open	Flat open	Flat open	Flat open
Number of template vertices (V)	1271	1089	1089	1189
Number of template triangles (T)	2400	2048	2048	2240
Video (vid.) or image collection (col.)	Vid.	Vid.	Vid.	Vid.
Number of images (N)	14 (68)	63 (313)	33 (100)	24 (71)
Image resolution ($w \times h$)	720 × 576	640 × 480	640 × 480	720 × 576
Correspondences per image N	1393	367	1228	4665
Focal length (px)	879.6	528.0	528.0	879.6
Focal length (% of w)	122.2	82.5	82.5	122.2
Lens opening angle (°)	44.5	62.4	62.4	44.5
Has ground truth 3D	No	Yes	Yes	No

6 Additional Initialization Sensitivity Experiments

6.1 Initialization Policies

We test 8 initialization policies in this experiment. The first 3 policies are the same as defined previously, and we introduce 5 new policies as follows:

- Policy 1: $\Psi_{init} = \{\psi^{An}\}$, $\mathcal{M} \in \{\text{MDH, PnP}\}$.
- Policy 2: $\Psi_{init} = \{20, 50, 80\}$, $\mathcal{M} \in \{\text{MDH, PnP}\}$.
- Policy 3: $\Psi_{init} = \{20, 30, 40, 50, 60, 70, 80\}$, $\mathcal{M} \in \{\text{MDH, PnP}\}$.
- Policy 4: $\Psi_{init} = \{50\}$, $\mathcal{M} \in \{\text{MDH}\}$.
- Policy 5: $\Psi_{init} = \{50\}$, $\mathcal{M} \in \{\text{MDH, PnP}\}$.
- Policy 6: $\Psi_{init} = \{\psi^{An}\}$, $\mathcal{M} \in \{\text{MDH}\}$.
- Policy 7: $\Psi_{init} = \{20, 50, 80\}$, $\mathcal{M} \in \{\text{MDH}\}$.
- Policy 8: $\Psi_{init} = \{\psi^{GT}\}$, $\mathcal{M} \in \{\text{MDH, PnP}\}$.

Policies 4 and 5 have one focal length sample whose opening angle is 50°. We compare them to evaluate the benefit of initializing with two SfT methods (MDH and PnP) compared with one (MDH). This is similarly done with policies 6 and 1, and policies 7 and 2.

Policies 6 and 1 have one focal length sample that is from the analytical method. Policies 7 and 2 have three focal length samples, and policy 3 has 7 focal length samples. Policy 8 has one focal length sample, which is the ground truth with opening angle denoted by ψ^{GT}. Of course, we cannot use policy 8 in practice because it requires the ground truth. However, we use it to compare how well the other policies perform compared to the ideal of initializing with the ground truth focal length.

6.2 Dataset Versions

We use six dataset versions in this experiment as follows:

- **v1**: No augmentation (original datasets)
- **v2**: Zoom augmentation and noise augmentation with $\sigma = 0.16\,w$
- **v3**: Zoom augmentation and noise augmentation with $\sigma = 0.32\,w$
- **v1 + SF**: v1 with Solvable Filtering
- **v2 + SF**: v2 with Solvable Filtering
- **v3 + SF**: v3 with Solvable Filtering

We now describe zoom augmentation, noise augmentation, and Solvable Filtering.

Zoom Augmentation Implementation
Zoom augmentation is implemented as follows. For each image in each dataset, we convert the point correspondences into retina coordinates, and then we projected them back to image coordinates using a simulated intrinsic matrix with a random focal length f_{rand}, with principal point at the image center and zero skew. We compute f_{rand} independently for each image, using an opening angle ψ_{rand} drawn with uniform probability in the range 10° to 90°, producing a wide range of focal lengths. We illustrate examples of images with simulated digital zoom for the Cap dataset in Fig. 12.

Noise Augmentation Implementation
Noise augmentation is implemented to simulate increasingly adverse conditions, by adding noise to each point correspondence and by reducing the number of point correspondences. Reducing points is required because additional noise has a smaller

Fig. 12 8 representative images from the Cap dataset where zoom augmentation is applied

influence on accuracy when there are many points. For each image in each dataset, we retain N' points sampled randomly and without replacement where N' is drawn uniformly in the range $[\min(N, 75), \min(N, 100)]$, where N is the original number of points in the image. Noise is added by randomly perturbing each of the retained image points by Gaussian I.I.D. noise of standard deviation σ (px). We test $\sigma = 0.16w$ and $\sigma = 0.32w$, where w is the image width. These are equivalent to standard deviations of 1px and 2px, respectively, at 640×480 resolution. The latter can be considered strong noise.

Solvable Filtering (SF) Implementation

There may exist problem instances that are not solvable by *any* fSfT method. This is a limitation because such instances dilute the effect of different initialization policies on the performance metrics. To deal with this, we also measure performance on a subset of problem instances for which the optimization-based method succeeds (we define success if the FLPE is below 15%). To implement this, we run the optimization-based method using *all* initializations contained in policies 1–8, and we filter out the problem instance if its corresponding FLPE was above 15%. By only evaluating on the filtered set of problem instances, we could answer the question: How well does an initialization policy perform given that the problem is solvable using at least one of the initialization policies?

6.3 Results

Focal length results are shown in Fig. 13, where Fig. 13a shows FLPE-success@15 and Fig. 13b shows FLPE-success@5 averaged across all datasets. We make the following observations:

17. There is a strong trend where using more focal length samples in the initialization policy reduces focal length error. There is a slight improvement using policy 3 (with 7 samples) compared to policy 2 (with 3 samples). By contrast, there is a strong benefit using policy 2 compared to policy 5 (with one sample). This illustrates diminishing returns where increasing the number of focal length samples has less of a benefit on solution accuracy.
18. The benefit of more focal length samples is less in v1 + SF compared to v2 + SF and v3 + SF. This is because in v1 + SF we do not apply zoom augmentation, so the focal lengths in v1 + SF have opening angles in the range $24.8° \leq \psi \leq 65.3°$. In these cases, the benefits of using more focal length samples are less pronounced compared to one sample at 50°.
19. There is a strong trend where using two SfT methods for initialization (MDH and PnP) improves performance compared to one SfT method (MDH). Recall that these methods operate very differently: MDH estimates deformation, and although it works well in general, there are cases when it does not estimate shape well thanks to the convex relaxation. By contrast, PnP does not estimate deformation, so the initialization it provides is the rigid pose that best fits the

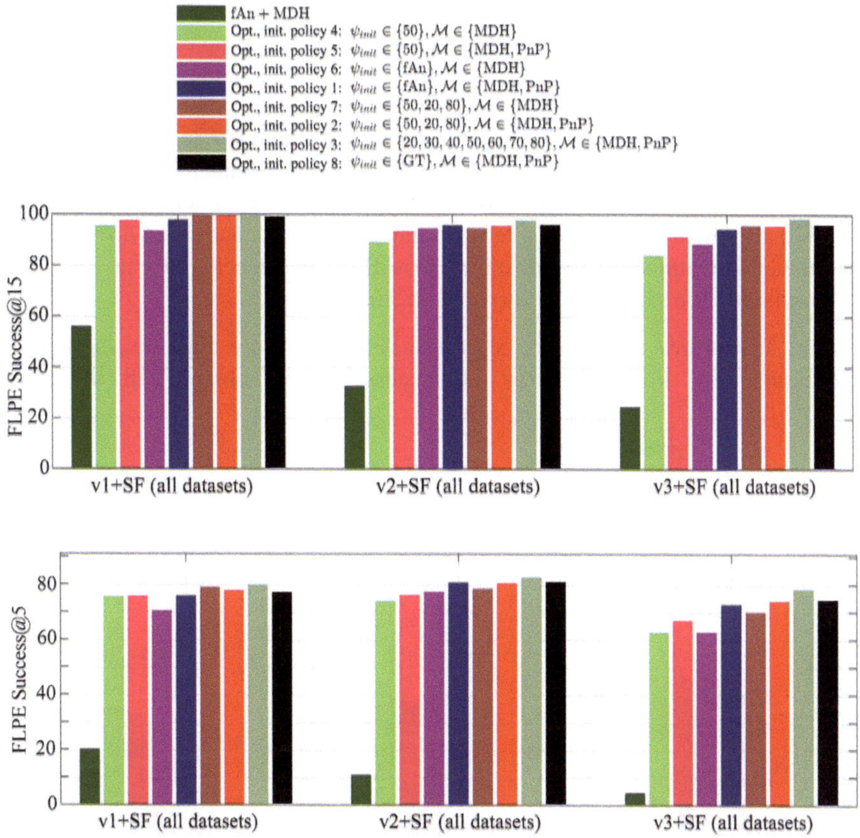

Fig. 13 Focal length percentage error (FLPE) performance of the analytical method (denoted as "fAn + MDH") and optimization-based method (denoted as "Opt.") using different initialization policies. The initialization policies are defined in terms of the set ψ_{init} of initial focal lengths and the set of SfT methods $\mathcal{M}$ used to initialize deformation

data. Adding the PnP solution appears to improve robustness in cases when the MDH solution cannot give a good initial estimate.

20. Initializing with the analytical method (policy 1) performs worse than initializing with a fixed opening angle of 50° (policy 5) for v1 + SF. However, for v2 + SF and v3 + SF, we see a benefit where policies 6 and 1 outperform policies 4 and 5. Recall that v2 + SF has zoom augmentation, and it has a much larger variation in opening angles compared to the original datasets without zoom augmentation (v1 + SF). Therefore, when there is larger variation in opening angles, the analytical method is able to provide a better initialization compared to using a fixed opening angle of 50°. By contrast, in v1 + SF, where the range of possible opening angles spans 40.5° with a midpoint at 45.0°, using a single opening angle of 50° performs better than using the opening angle from the analytical method.

21. Initializing with the ground truth focal length (policy 8) performs approximately the same as policy 2 (three focal length samples) for all dataset versions. This shows that accurate focal length initialization is not required by Algorithm 1.
22. Initializing with policy 8 performs worse in general than initializing with policy 3 (7 focal length samples). This can seem counter intuitive, and we study the cause in more detail below. In short, the reason is because when we initialize with multiple focal length samples, we introduce shape diversity into the initialization set as a side effect. This diversity can help to locate the global optimum. We call this the *diverse initialization effect.*

The shape error results are shown in Fig. 14, where Fig. 14a shows SE-success@5 and Fig. 14b shows SE-success@2. We observe all the same

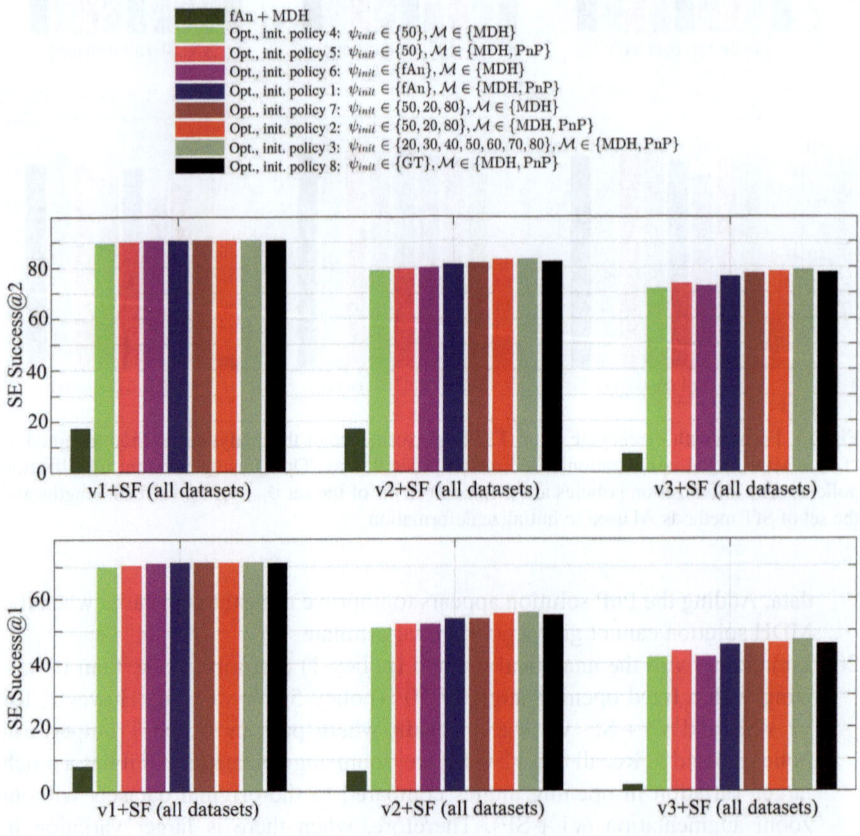

Fig. 14 Shape error (SE) performance of the analytical method (denoted as "fAn + MDH") and optimization-based method (denoted as "Opt.") using different initialization policies. The initialization policies are defined in terms of the set ψ_{init} of initial focal lengths and the set of SfT methods $\mathcal{M}$ used to initialize deformation

performance trends as we have observed for FLPE. The diverse initialization effect is also present.

7 Computation Cost Analysis

We compare the computational cost of the different initialization policies by measuring the average number of optimization iterations (Gauss–Newton steps) required by Algorithm 1 using each policy. We use this instead of computation time because it is invariant to the implementation platform, and it is roughly proportional to computation time because the cost of executing each Gauss–Newton iteration is approximately constant at each iteration. The results are shown in Fig. 15 where we observe the following:

23. There is a clear increase in computational cost using policies with a larger initialization set.
24. There is practically no difference in the computational cost of initializing using one focal length sample (policies 4 and 5) and using the analytical method's focal length estimate (policies 6 and 1). This indicates that the number of iterations required for convergence is not highly sensitive to the accuracy of the initial focal length estimate.
25. The early termination criteria used in Algorithm 1 to avoid repeated search of solution space are proving effective. Without them, we would be seeing a doubling in the number of optimization iterations from policies using MDH

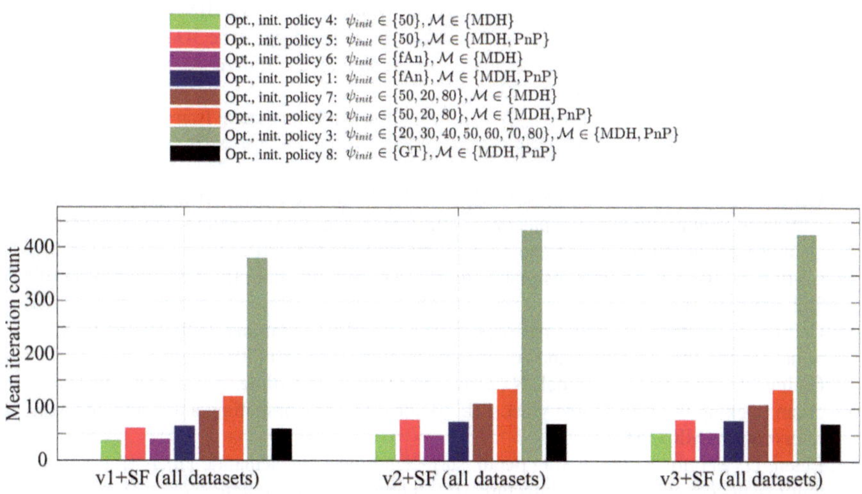

Fig. 15 Computational cost of the optimization-based method with different initialization policies. This is expressed in the average number of Gauss–Newton iterations required for Algorithm 1 to converge

to policies using both MDH and PnP. For example, the extra cost from policy 7 to policy 2 is between 22.2% and 32.7% depending on the dataset version. Without early termination, the additional cost would be approximately 100%.

26. There is a slight increase in computational cost from v1 to v2 (and v1 + SF to v2 + SF) for all policies. This indicates that increasing noise also increases the number of iterations required for convergence.

References

1. Bansal, A., Russell, B., Gupta, A.: Marr revisited: 2D-3D alignment via surface normal prediction. In: 2016 IEEE Conference on Computer Vision and Pattern Recognition (CVPR), Las Vegas, NV, USA, 2016. https://doi.org/10.1109/CVPR.2016.642
2. Bartoli, A., Collins, T.: Template-based isometric deformable 3d reconstruction with sampling-based focal length self-calibration. In: International Conference on Computer Vision and Pattern Recognition (CVPR), pp. 1514–1521 (2013)
3. Bartoli, A., Pizarro, D., Collins, T.: A robust analytical solution to isometric shape-from-template with focal length calibration. In: International Conference on Computer Vision (ICCV) (2013)
4. Bartoli, A., Gérard, Y., Chadebecq, F., Collins, T., Pizarro, D.: Shape-from-template. IEEE IEEE Trans. Pattern Anal. Mach. Intell. **37**(10), 2099–2118 (2015)
5. Brunet, F., Hartley, R., Bartoli, A., Navab, N., Malgouyres, R.: Monocular template-based reconstruction of smooth and inextensible surfaces. In: Asian Conference on Computer Vision (ACCV) (2010)
6. Brunet, F., Hartley, R., Bartoli, A.: Monocular template-based 3D surface reconstruction: convex inextensible and nonconvex isometric methods. Comput. Vis. Image Underst. **125**, 138–154 (2014)
7. Chhatkuli, A., Pizarro, D., Bartoli, A.: Non-rigid shape-from-motion for isometric surfaces using infinitesimal planarity. In: British Machine Vision Conference (BMVC) (2014)
8. Chhatkuli, A., Pizarro, D., Bartoli, A.: Stable template-based isometric 3D reconstruction in all imaging conditions by linear least-squares. In: International Conference on Computer Vision and Pattern Recognition (CVPR) (2014)
9. Chhatkuli, A., Pizarro, D., Bartoli, A., Collins, T.: A stable analytical framework for isometric shape-from-template by surface integration. IEEE Trans. Pattern Anal. Mach. Intell. **39**(5), 833–850 (2017)
10. Collins, T., Durou, J.-D., Gurdjos, P., Bartoli, A.: Single view perspective shape-from-texture with focal length estimation: a piecewise affine approach. In: International Symposium on 3D Data Processing, Visualization and Transmission (3DPVT) (2010)
11. Collins, T., Bartoli, A.: Infinitesimal plane-based pose estimation. Int. J. Comput. Vis. **109**(3), 252–286 (2014)
12. Collins, T., Bartoli, A.: Realtime shape-from-template: system and applications. In: International Symposium on Mixed and Augmented Reality (ISMAR) (2015)
13. Collins, T., Bartoli, A., Bourdel, N., Canis, M.: Robust, real-time, dense and deformable 3D organ tracking in laparoscopic videos. In: Medical Image Computing and Computer-Assisted Intervention (MICCAI) (2016)
14. Eigen, D., Fergus, R.: Predicting depth, surface normals and semantic labels with a common multi-scale convolutional architecture. In: International Conference on Computer Vision (ICCV), pp. 2650–2658 (2015)
15. Fuentes-Jimenez, D., Casillas-Perez, D., Pizarro, D., Collins, T., Bartoli, A.: Deep shape-from-template: wide-baseline, dense and fast registration and deformable reconstruction from a single image (2018). arXiv:1811.07791

16. Fuentes-Jimenez, D., Pizarro, D., Casillas-Perez, D., Collins, T., Bartoli, A.: Texture-generic deep shape-from-template. IEEE Access **9**, 75211–75230 (2021)
17. Gallardo, M., Collins, T., Bartoli, A.: Can we jointly register and reconstruct creased surfaces by shape-from-template accurately? In: European Conference on Computer Vision (ECCV) (2016)
18. Gallardo, M., Collins, T., Bartoli, A.: Dense non-rigid structure-from-motion and shading with unknown albedos. In: International Conference on Computer Vision (ICCV) (2017)
19. Garg, R., Kumar, B.V., Carneiro, G., Reid, I.: Unsupervised cnn for single view depth estimation: geometry to the rescue. In: European Conference on Computer Vision (ECCV) (2016)
20. Golyanik, V., Shimada, S., Varanasi, K., Stricker, D.: HDM-Net: monocular non-rigid 3D reconstruction with learned deformation model. In: Bourdot, P., Cobb, S., Interrante, V., kato, H., Stricker, D. (eds.) Virtual Reality and Augmented Reality. EuroVR 2018. Lecture Notes in Computer Science, vol. 11162. Springer, Cham (2018). https://doi.org/10.1007/978-3-030-01790-3_4
21. Güler, R.A., Neverova, N., Kokkinos, I.: Densepose: dense human pose estimation in the wild. In: International Conference on Computer Vision and Pattern Recognition (CVPR), pp. 7297–7306 (2018)
22. Ilic, S., Salzmann, M., Fua, P.: Implicit meshes for effective silhouette handling. Int. J. Comput. Vis. **72**, 159–178 (2007)
23. Ke, T., Roumeliotis, S.I.: An efficient algebraic solution to the perspective-three-point problem. In: International Conference on Computer Vision and Pattern Recognition (CVPR), pp. 4618–4626 (2017)
24. Kimmel, R., Sethian, J.: Computing geodesic paths on manifolds. Proc. Natl. Acad. Sci. U. S. A. **95**, 8431–8435 (1998)
25. Koo, B., Özgür, E., Le Roy, B., Buc, E., Bartoli, A.: Deformable registration of a preoperative 3D liver volume to a laparoscopy image using contour and shading cues. In: Medical Image Computing and Computer-Assisted Intervention (MICCAI) (2017)
26. Levi, Z., Gotsman, C.: Smooth rotation enhanced as-rigid-as-possible mesh animation. IEEE Trans. Vis. Comput. Graph. **21**(2), 264–277 (2015)
27. Liu, F., Shen, C., Lin, G., Reid, I.: Learning depth from single monocular images using deep convolutional neural fields. IEEE Trans. Pattern Anal. Mach. Intell. **38**(10), 2024–2039 (2016). https://doi.org/10.1109/TPAMI.2015.2505283
28. Liu-Yin, Q., Yu, R., Agapito, L., Fitzgibbon, A., Russell, C.: Better together: joint reasoning for non-rigid 3D reconstruction with specularities and shading. In: British Machine Vision Conference (BMVC) (2016)
29. Lowe, D.G.: Distinctive image features from scale-invariant keypoints. Int. J. Comput. Vision **60**(2), 91–110 (2004)
30. Magnenat, S., Ngo, D., Zünd, F., Ryffel, M., Noris, G., Roethlin, G., Marra, A., Nitti, M., Fua, P., Gross, M., Sumner, R.W.: Live texturing of augmented reality characters from colored drawings. IEEE Trans. Vis. Comput. Graph. **21**, 1201–1210 (2015)
31. Malti, A., Bartoli, A., Collins, T.: A pixel-based approach to template-based monocular 3d reconstruction of deformable surfaces. In: International Conference on Computer Vision Workshops, pp. 1650–1657 (2011)
32. Martinez, J., Hossain, R., Romero, J., Little, J.J.: A simple yet effective baseline for 3d human pose estimation. In: International Conference on Computer Vision (ICCV) (2017)
33. Mikolajczyk, K., Tuytelaars, T., Schmid, C., Zisserman, A., Matas, J., Schaffalitzky, F., Kadir, T., Gool, L.V.: A comparison of affine region detectors. Int. J. Comput. Vis. **65**, 2005 (2005)
34. MOSEK ApS. The MOSEK optimization toolbox for MATLAB manual. Version 9.0. (2019)
35. Ngo, T.D., Park, S., Jorstad, A.A., Crivellaro, A., Yoo, C., Fua, P.: Dense image registration and deformable surface reconstruction in presence of occlusions and minimal texture. In: International Conference on Computer Vision (ICCV) (2015)
36. Ostlund, J., Varol, A., Ngo, T., Fua., P.: Laplacian meshes for monocular 3D shape recovery. In: European Conference on Computer Vision (ECCV) (2012)

37. Perriollat, M., Hartley, R., Bartoli, A.: Monocular template-based reconstruction of inextensible surfaces. Int. J. Comput. Vis. **95**(2), 124–137 (2011)
38. Pilet, J., Lepetit, V., Fua, P.: Fast non-rigid surface detection, registration and realistic augmentation. Int. J. Comput. Vis. **76**(2), 109–122 (2008)
39. Pizarro, D., Bartoli, A.: Feature-based deformable surface detection with self-occlusion reasoning. Int. J. Comput. Vis. **97**(1), 54–70 (2012)
40. Pumarola, A., Agudo, A., Porzi, L., Sanfeliu, A., Lepetit, V., Moreno-Noguer, F.: Geometry-aware network for non-rigid shape prediction from a single view. In: International Conference on Computer Vision and Pattern Recognition (CVPR), pp. 4681–4690. IEEE Computer Society, Washington (2018)
41. Salzmann, M., Fua, P.: Reconstructing sharply folding surfaces: a convex formulation. In: International Conference on Computer Vision and Pattern Recognition (CVPR), pp. 1054–1061 (2009)
42. Salzmann, M., Hartley, R., Fua, P.: Convex optimization for deformable surface 3D tracking. In: International Conference on Computer Vision (ICCV) (2007)
43. Salzmann, M., Moreno-Noguer, F., Lepetit, V., Fua, P.: Closed-form solution to non-rigid 3d surface registration. In: European Conference on Computer Vision (ECCV), pp. 581–594 (2008)
44. Salzmann, M., Urtasun, R., Fua, P.: Local deformation models for monocular 3D shape recovery. In: International Conference on Computer Vision and Pattern Recognition (CVPR) (2008)
45. Sattler, T., Sweeney, C., Pollefeys, M.: On sampling focal length values to solve the absolute pose problem. In: European Conference on Computer Vision (ECCV), Cham, pp. 828–843 (2014)
46. Schaefer, S., McPhail, T., Warren, J.: Image deformation using moving least squares. ACM Trans. Graph. **25**(3), 533–540 (2006)
47. Sorkine, O., Alexa, M.: As-rigid-as-possible surface modeling. In: Proceedings of the Fifth Eurographics Symposium on Geometry Processing, Aire-la-Ville, SGP '07, pp. 109–116. Eurographics Association, Eindhoven (2007)
48. Sturm, J.: Using SeDuMi 1.02, a MATLAB toolbox for optimization over symmetric cones. Optimization Methods and Software 11–12 (1999), pp. 625–653. Version 1.05 available from http://fewcal.kub.nl/sturm
49. Sturm, P.F., Maybank, S.J.: On plane-based camera calibration: a general algorithm, singularities, applications. In: International Conference on Computer Vision and Pattern Recognition (CVPR), vol. 1, pp. 432–437 (1999)
50. Terzopoulos, D., Platt, J., Barr, A., Fleischer, K.: Elastically deformable models. SIGGRAPH. Comput. Graph. **21**(4), 205–214 (1987)
51. Tomasi, C., Kanade, T.: Detection and Tracking of Point Features. Shape and Motion from Image Streams, School of Computer Science, Carnegie Mellon Univ. (1991)
52. Tran, Q.-H., Chin, T.-J., Carneiro, G., Brown, M.S., Suter, D.: In defence of RANSAC for outlier rejection in deformable registration. In: European Conference on Computer Vision (ECCV) (2012)
53. Varol, A., Salzmann, M., Fua, P., Urtasun, R.: A constrained latent variable model. In: International Conference on Computer Vision and Pattern Recognition (CVPR) (2012)
54. Vávra, P., Roman, J., Zonća, P., Ihnát, P., Némec, M., Jayant, K., Habib, N., El-Gendi, A.: Recent development of augmented reality in surgery: a review. J. Healthc. Eng. **2017**, 1–9 (2017)
55. Vicente, S., Agapito, L.: Balloon shapes: reconstructing and deforming objects with volume from images. In: International Conference on 3D Vision (2013)
56. Wang, X., Fouhey, D.F., Gupta, A.: Designing deep networks for surface normal estimation. In: International Conference on Computer Vision and Pattern Recognition (CVPR), pp. 539–547 (2015)

57. Yi, K.M., Trulls, E., Lepetit, V., Fua, P.: Lift: learned invariant feature transform. In: Leibe, B., Matas, J., Sebe, N., Welling, M. (eds.) Computer Vision – ECCV 2016, pp. 467–483. Springer International Publishing, Cham (2016)
58. Yu, R., Russell, C., Campbell, N.D.F., Agapito, L.: Direct, dense, and deformable: template-based non-rigid 3D reconstruction from RGB video. In: International Conference on Computer Vision (ICCV), pp. 918–926 (2015)
59. Zhang, Z.: A flexible new technique for camera calibration. IEEE Trans. Pattern Anal. Mach. Intell. **22**, 1330–1334 (2000)

Reconstruction of a Botanical Tree from a 3D Point Cloud

J. Andreas Bærentzen, Ida Bukh Villesen, and Ebba Dellwik

Abstract 3D models are often acquired using optical methods such as LiDAR, structured light, or automated photogrammetry. These methods produce point clouds, and the typical downstream processing pipeline consists of registration of individually scanned point clouds followed by reconstruction of a triangle mesh from the combined point cloud. In this paper we consider a specific challenge that might prevent this pipeline from producing meshes suitable for later applications. The challenge concerns reconstruction of 3D models with thin tubular features, here exemplified by a tree with a very complex crown structure, where the radii of some branches are on the same order as the sample distance. In such cases, traditional surface reconstruction methods perform poorly. We discuss how a surface can still be reconstructed from this type of data. Our procedure begins by constructing a skeleton of the object from a graph whose vertices are the input points, a surface representation is then created from the skeleton, and, finally, a triangle mesh is generated from the surface representation. We demonstrated the efficacy of our method on a tree acquired using ground-based LiDAR.

Keywords Scanning · Graph · Point cloud · Reconstruction · Registration · Alignment

---

J. A. Bærentzen (✉) · I. B. Villesen
Department of Applied Mathematics and Computer Science, Technical University of Denmark, Lyngby, Denmark
e-mail: janba@dtu.dk; s174385@student.dtu.dk

E. Dellwik
Department of Wind Energy, Technical University of Denmark, Roskilde, Denmark
e-mail: ebde@dtu.dk

E. Cristiani et al. (eds.), *Mathematical Methods for Objects Reconstruction*, Springer INdAM Series 54, https://doi.org/10.1007/978-981-99-0776-2_4

1 Introduction

For a wide range of purposes, we require 3D models of real-world objects, and a great many methods for obtaining such models have been devised. These methods range from coordinate measuring machines, over laser scanners, structured light scanners, to reconstruction using photogrammetry. Regardless of whether the methods are mechanical or (more likely) optical, they almost invariably produce an output in the form of a 3D point cloud, and downstream processing is then used to turn this cloud into a surface—often in the form of a triangle mesh. This is a much studied problem, and Berger et al. provide a comprehensive survey [1] of methods for surface reconstruction from point clouds.

Unfortunately, surface reconstruction methods are typically designed to handle a limited range of scales. This leads to challenges in the case of many botanical objects such as trees which tend to have features at different scales that are often orders of magnitude apart. This type of data is the main concern in this paper where we propose a pipeline for reconstruction of botanical trees and demonstrate it on a LiDAR scanning of an oak tree.

1.1 Context and Previous Works

Popular methods for reconstruction from point clouds include volumetric methods such as Poisson Reconstruction [2] and combinatorial methods such as the ball pivot algorithm [3, 4]. In many cases, we obtain a good result using one of these approaches, but we sometimes scan objects with features that are too fine for the reconstruction process. A good reconstruction requires that the distance from the surface to the medial surface of the object we mean to reconstruct is large compared to the distance between adjacent point samples. The medial surface is the locus for which the closest surface point is not unique [5]. For example, for a cylinder, the medial surface degenerates to a curve which is simply its axis, and the distance from the surface to this axis is obviously the radius of the cylinder. Constraints on sample spacing can now be expressed relative to the smallest distance from the surface to its medial surface. For instance, Amenta et al. [6] require that the distance to the medial surface be at least twice the distance to the closest sample from any point on the surface.

On the other hand, even if the sampling is not sufficiently dense to reliably reconstruct a surface in the case of tubular structures, it might be feasible to reconstruct a curve skeleton. A curve skeleton is a graph, which approximates the centerlines of the tubular structures from the point cloud, and from the curve skeleton, we can create an implicit surface description, e.g., using a method known as convolutional surfaces [7], which is the approach taken in this paper.

One advantage of using an implicit representation is that it ensures a watertight mesh. A mesh is deemed watertight if we can distinguish the interior from the exterior. Specifically, any path connecting a point considered interior relative to the surface to a point considered exterior must intersect the surface an odd number of times. An application that requires water-tightness is wind simulation. In order to achieve an accurate result, the boundary conditions should be represented as solid watertight surfaces, see, e.g., [8].

Previous works on the reconstruction of trees from point clouds include the studies by Livny et al. [9], Raumonen et al. [10], and Hackenberg et al. [11]. Livny et al demonstrated a successful reconstruction of urban trees using sparse scans using a series of algorithms including a skeletonization of the tree structure. However, for highly complex and dense crown structure, the low scanning density did not admit a realistic reproduction. To overcome the limitation of the low-density scans, they applied established fixed allometric functions [12] describing the relationships between the finer (daughter) and the coarser (mother) branches. Since Raumomen et al. and Hackenberg et al. used higher density scans, the surfaces could be reconstructed from the scans more directly by fitting different diameter cylinders to the branches. Both these studies were focused on the estimation of wood volume, and water-tightness was not in focus of their works.

Here, we present a set of algorithms, which like in Livny et al. are based on skeletonization. Due to a higher scanning density in the original point cloud, the method also allows for a different and more realistic way of estimating the mother-daughter diameter relationship which builds on the classical Da Vinci relationship for trees [13].

1.2 Materials

The study is focused on the open-grown oak tree shown in Fig. 1. This tree, which is located near the shore of the Roskilde Fjord in Denmark, has been the focus point in a research project which had the focus to improve the understanding of wind-tree

Fig. 1 On the left an open-grown oak tree which was scanned and reconstructed using the methods discussed in this paper. The image on the right shows the initial LiDAR scanned point cloud

interaction. The exact location of the tree and a description of the core experiment can be found in [14, 15]. The tree was scanned in April 2017 using a Leica Nova MS60 LiDAR scanner. The leaves of the tree had not yet come out and, hence, the tree is a perfect example of a collection of thin tubular structures of a wide range of diameters in a complex crown structure. The tree was scanned from six angles at 3–10 m distance from the tree, on a day with no wind. Each scan was taken with a resolution of 2 cm. In order to make sure that the point cloud from each scan could be merged into the same coordinate system, the location and orientation of the scanner for each of the scans was determined precisely by measuring the exact location of 1 cm wide marker near the top of four sticks, placed 30–50 m away from the tree. Having removed background points, the tree itself was represented as 832,943 points Fig. 1(left). Compared to the literature mentioned above, this point density is less than (but of similar magnitude) the density used in [10], it is an order of magnitude fewer points than in [11], and it is 1–2 order of magnitudes more points compared to the highly undersampled trees in [9].

2 From Graphs of Point Clouds to Tubular Surfaces

Our procedure can roughly be divided into three phases:

- Initially, a graph is constructed by connecting each point to its nearby neighbors within a given distance as described in Sect. 2.1. The graph is further augmented and perhaps simplified as discussed in Sect. 2.2. Both of these steps aid the following skeletonization step.
- Next, we find a skeletal structure which approximates the centerline of the (tubular) surface that was scanned in the first place. The local separator (LS) method for skeletonization is discussed in Sect. 2.3.
- Finally, we ensure that the skeleton connectivity is correct by reattaching skeletal branches which are disconnected from the rest of the structure and by cutting loops to ensure that the tree skeleton is also a tree in the graph-theoretical sense. These steps are covered in Sect. 2.4.1. The last part of the pipeline in Sect. 2.4.2 concerns estimating the branch thicknesses and performing the final surface reconstruction.

The different phases as well as the different algorithms used in each phase are shown in Fig. 2.

2.1 Constructing Graphs from Point Clouds

Before demonstrating the new algorithms on the point cloud, we use a synthetic example to demonstrate the pipeline for reconstructing an object with thin tubular

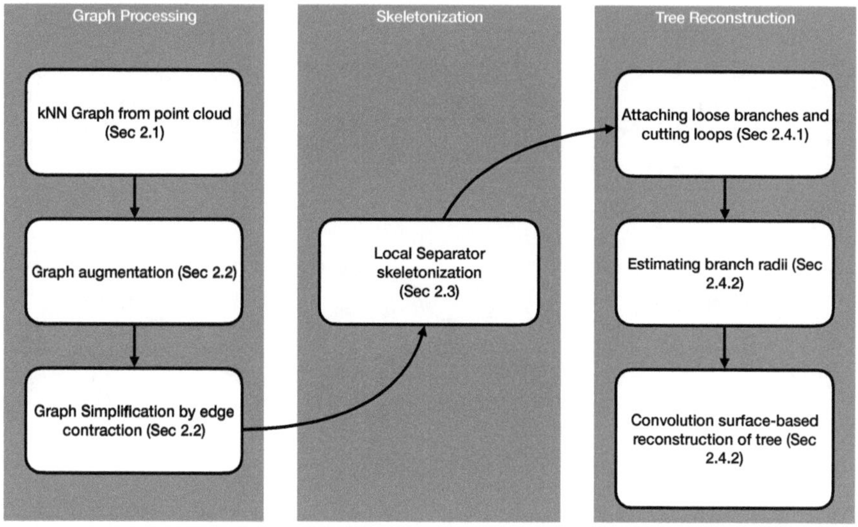

Fig. 2 A visual overview of our procedure which consists of three main parts. First a graph processing step, next skeletonization, and finally reconstruction of the tree from its skeleton

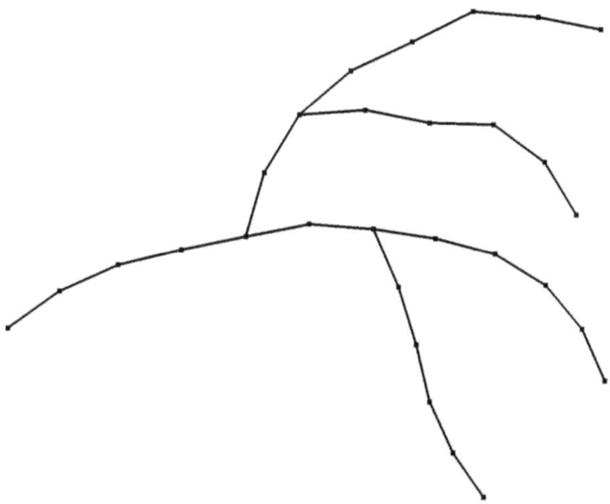

Fig. 3 A simple branching structure which serves as a synthetic example which we use to demonstrate the challenges in reconstructing the correct connectivity of a graph from a set of noisy samples

features. The example, shown in Fig. 3, is a simple tree structure with three bifurcations.

In order to create a synthetic point set, we sampled ten random points on each of the 28 edges (of length 2) and created a point for each sample by adding a random

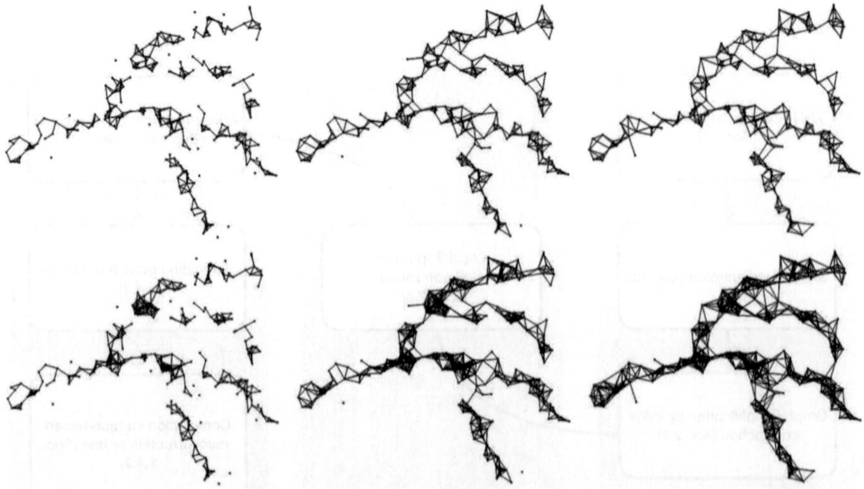

Fig. 4 Noisy samples from the synthetic graph have been connected in order to form new graphs. From left to right, points at distances smaller than a threshold of 0.9, 1.2, and 1.5 are connected. In the top row, points are connected to their four nearest neighbors closer than this threshold, and in the bottom row points are connected to their nine nearest neighbors closer than the threshold

vector $\mathbf{v} = [\mathcal{N}(0, 0.45), \mathcal{N}(0, 0.45), \mathcal{N}(0, 0.45)]^T$ where $\mathcal{N}(0, 0.45)$ denotes a normal distribution of zero mean and variance 0.45.

A 2×3 matrix of graphs, $G_{k;d}$, is then created by connecting these points to their k neighbors within a distance d, where $k \in \{4, 9\}$ and $d \in \{0.9, 1.2, 1.5\}$. The graphs are shown in Fig. 4. The tradeoffs are very clear: if we choose a distance threshold that is too small, we obtain a disconnected graph: 0.9 is clearly not sufficient to close the gap between points that should be connected as observe in the leftmost column. However, choosing the distance threshold too large results in spurious connections as evident in the rightmost column. Especially for $k = 9$ there is an unfortunate connection between the two branches.

Summing up,

- d must be smaller than the distance between features, but two tubular features may be closer than the average sample distance in which case the features cannot be separated.
- d must be greater than the distance between adjacent samples on the same surface component to ensure that such samples are connected.

In our synthetic example, the value $d = 1.2$ is observed to give the best result. The method we will discuss in the following is less sensitive to the value of k. However, if k is too small, we again risk that structures are disconnected.

2.2 Graph Augmentation and Simplification

Unfortunately, while $G_{9;1.2}$ visually captures the structure of the original graph (shown in Fig. 3), directly skeletonizing this graph using the LS method, which will be discussed in the sequel, does not produce a good result as shown in Fig. 5(top row). Essentially, the problem is that the connectivity of $G_{9;1.2}$ is too low because *chordless cycles* in the graph also become cycles in the skeleton. In a chordless cycle none of the adjoined vertices is connected by an edge that does not itself belong to the cycle.

Fortunately, we can reduce the number of chordless cycles by a very simple augmentation scheme. Let $N(u)$ be the set of neighbors of vertex u, we can now define $N^k(u)$ as k'th order neighbors: all vertices separated from u by up to k edges. Now, we simply add an edge between any pair of vertices $u, v \in G_{9;1.2}$ if $v \in N^k(u)$. The LS skeleton produced from the resulting augmented graph $\widetilde{G}_{9;1.2}$ does not contain any spurious cycles as shown in Fig. 5. In fact, the branching structure perfectly matches the ground truth even if the geometry is not exactly the same. However, with the added level noise, some distortion is expected.

The synthetic example is very simple. For graphs of greater complexity, it may be beneficial to first simplify the graph. Many graphs can be simplified quite significantly with only limited changes to the skeleton ultimately produced.

Graph simplification can be performed by edge contraction. When an edge is contracted, two vertices are moved to their barycenter, and one of the vertices is removed. The surviving vertex inherits all edges that were previously incident on the one that was removed. It is important to avoid contracting in such a way that a very unbalanced graph results. For instance, if many vertices are contracted to just

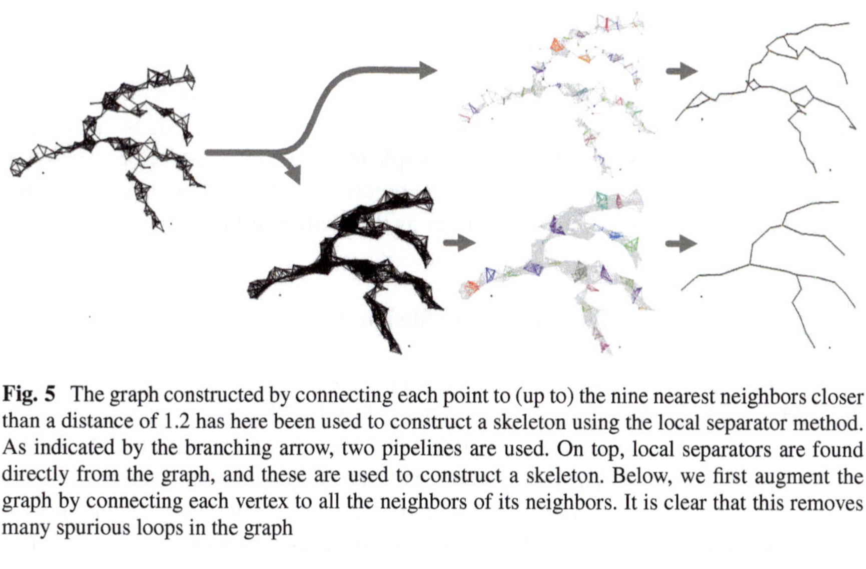

Fig. 5 The graph constructed by connecting each point to (up to) the nine nearest neighbors closer than a distance of 1.2 has here been used to construct a skeleton using the local separator method. As indicated by the branching arrow, two pipelines are used. On top, local separators are found directly from the graph, and these are used to construct a skeleton. Below, we first augment the graph by connecting each vertex to all the neighbors of its neighbors. It is clear that this removes many spurious loops in the graph

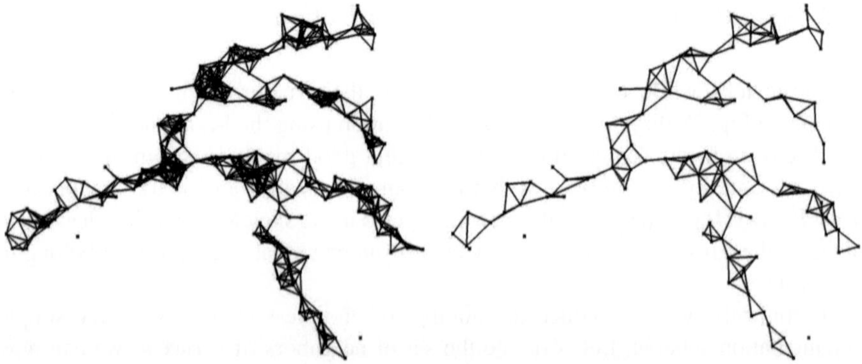

Fig. 6 On the left, $G_{9;1.2}$ is shown and on the right the same graph has been simplified by contracting edges shorter than 0.75

a few vertices these vertices will end up having a very high valence. Two principles seem to suffice to ensure a balanced contraction:

- Contractions are performed in order of increasing edge length. The process stops when no edges are shorter than a set threshold.
- The contraction process performs a sweep of the entire graph in each iteration. If a vertex is the result of a contraction performed in a given sweep then no edge incident on that vertex is contracted later in the same sweep.

An example of the application of our contraction procedure is illustrated in Fig. 6.

2.3 *The Local Separator Algorithm*

The only remaining part of the pipeline used to reconstruct the synthetic skeleton in Fig. 3 is the Local Separator (LS) algorithm for skeletonization. This is the topic of Bærentzen and Rotenberg [16], but for completeness we give a brief description in the following. However, first we need to be precise about the meaning of a skeleton which is a bit harder to define as precisely as the medial surface.

2.3.1 The Medial Surface and Curve Skeletons

The *medial surface* of a shape is the locus of points for which the closest point on the surface is not unique [5, 17]. As the name implies, this locus is not necessarily a curve. In fact it is usually a collection of connected surface patches. However, the definition of the medial surface is precise, and if we know the distance for each point belonging to the medial surface of an object, we can in theory reconstruct the object from this representation. In comparison, the notion of a *curve skeleton* is less precise. A curve skeleton is generally understood to be a collection of curves that are roughly centerlines of features of the object [18]. Some authors define curve skeletons to lie in the medial surface [19], but this is not universal. Like the medial surface, a curve skeleton captures a lot of information about both the geometry and the topology of the object, but it is not possible to reconstruct the object from a curve skeleton. However, as we shall see, it is possible to create an algorithm that finds the curve skeleton from a graph created from point samples. While some authors have proposed methods for computing the medial surface from a point sampling of the surface [20], invertibility of the medial surface transform seems to imply that the surface must be sampled in all areas where we wish to compute the medial surface. However, curve skeletonization is not invertible, and, hence, more forgiving.

Generally, algorithms for curve skeletonization also take a surface representations as input, and, furthermore, most methods for curve skeletonization are informed by homotopy meaning that the skeleton is found by contracting the surface [21]. Clearly, we cannot use a method that works only for surfaces since the neighborhood graphs considered in this paper are very far from qualifying. In principle, we could still employ a contraction based method which does not require a surface representation as input. On the other hand, the successful contraction based method by Tagliasacchi et al. [21] still relies on surface information in order to keep the skeleton centered. Moreover, this type of method has a tendency to remove salient features simply because contraction is based on smoothing which means that small details are sometimes smoothed away before becoming part of the skeleton.

2.3.2 Local Separators

In graph theory, vertex separators are sets of vertices such that the graph would break into two disconnected parts if they were removed. Many separators are not helpful. For any vertex, its neighbors are clearly a separator which separates that vertex from the rest of the graph. Moreover, if we think of a torus we might have a set of vertices whose removal would turn the torus into a cylinder (topologically), but since that does not change the number of components, it is not formally a separator. Hence, we introduce the notion of a *local separator*. Local separators are separators of an induced subgraph, and a ring of vertices which cuts a torus *does* qualify as a local separator [16].

The local separator algorithm operates by seeding an expanding front at a given vertex. Vertices are iteratively added to front; when the front meets itself, a local separator has been found, and vertices are subsequently removed until the separator is minimal. This procedure leads to a super set of local separators, and a set packing strategy is subsequently employed to find a maximal disjoint collection of local separators [22].

2.3.3 Skeleton Construction

The skeleton is produced from the packed local separators using a simple two step procedure. First, we recursively assign vertices to their closest separator. Subsequently, a skeletal node is created for each separator, and its geometric position is the barycenter of the vertices in the separator. For each pair of adjacent separators, the corresponding skeletal vertices are connected, and where three or more separators are mutually adjacent, we introduce a branch vertex connecting to all mutually adjacent separators. We refer the reader to [16] for more details about all aspects of the local separator algorithm.

A benefit of the local separator algorithm is that it operates on graphs whose vertices map to geometric positions, and the graphs do not have to be planar or manifold embeddable. In Fig. 7 we illustrate this point by showing separators on graphs that are produced from three very different sources, namely

- A binary voxel grid where the interior voxels constitute the graph vertices and each vertex is connected to its 26 neighbors.
- A triangle mesh which is clearly already a graph.
- A synthetically generated graph.

Figure 5 shows both separators and the final skeletons for the synthetic graph that we have been using to illustrate the entire pipeline. It may be slightly counterintuitive that the augmented graph contains much fewer separators and has a simpler skeleton than the non-augmented graph, but the more connected a graph becomes, the fewer separators it contains. In the extreme case of a fully connected graph (where all pairs of vertices are connected) there are no separators—neither local nor global.

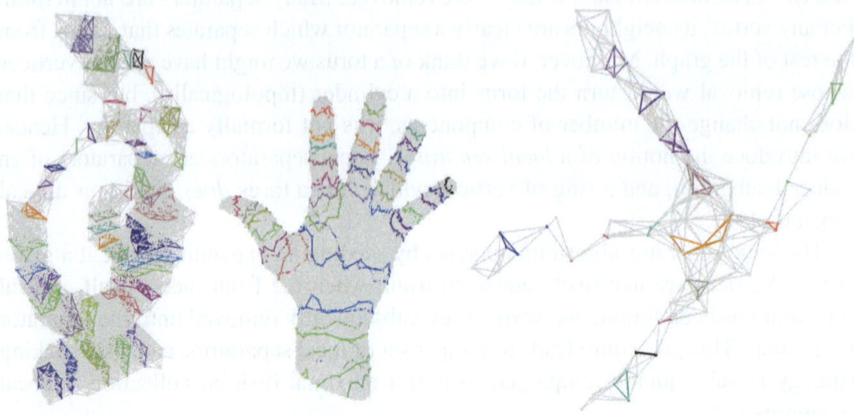

Fig. 7 From left to right: a graph formed from a voxel grid, a triangle mesh, and a synthetically generated graph. On all three graphs, each local separator is shown in a unique color

2.4 *Botanical Tree*

At this point, we have a procedure that allows us to compute a skeleton from a collection of points connected to form a graph even if a surface reconstruction is not possible. This procedure was applied to the point cloud described in Sect. 1.2.

To produce the graph for skeletonization, we applied the procedure as described in the previous sections with the following parameters

- The original points were converted to vertices of a graph and two vertices were connected if closer than $d = 2\,\text{cm}$. A point was connected to at most $k = 15$ neighbors.
- We augmented the graph significantly by connecting each vertex, u, to all vertices in $N^5(u)$ but only within a radius of 5 cm from u.
- Edges shorter than 2.5 cm were contracted.
- The skeleton was computed producing a new (skeletal) graph that can be understood as a model of the structure of the tree.

This, however, does not mark the end of the pipeline. Even though the scanner captured a lot of the tree and most of the captured branches are represented in the resulting skeleton, some of the branches are not connected to the rest of the tree. An estimated 16.7% of the skeleton was not connected to the rest to be precise [23]. Moreover, we wanted to reconstruct the surface of the tree which involved estimating the thickness of the branches. These two processes are discussed in the following.

2.4.1 Reconnecting the Skeleton

In order to ensure that the graph-theoretical tree is a reasonable skeleton for a botanical tree, we need to reattach any branches that are not connected to the main part of the skeleton. For the following processes to work, we need to assume that a root vertex has been identified. This vertex is simply the skeleton vertex closest to the ground (i.e., lowest z value).

The process of reattachment begins with the identification of so-called critical vertices on the disconnected subgraphs (smaller branches). These are the vertices which the algorithm will try to attach to the main part of the skeleton (the actual tree). The following heuristic is used to identify critical points.

- If a disconnected subgraph contains no nodes of valence >2 (i.e., branch vertices), the critical points are the end points (of valence 1) as well as vertices where the angle between the two incident edges is less than 130°.
- If a disconnected subgraph does contain nodes of valence >2, the angle threshold is set to 90° and both vertices of valence 1 or valence >2 are now considered critical.

Once the critical points have been identified, their attachment points are found among a set of the closest points on the main part of the skeleton. For subgraphs without vertices of valence >2, we also prioritize that the connection is straight since these are likely continuations of branches already present in the main graph. For subgraphs with vertices of valence >2, we also prioritize connection points that have a short path to the root of the tree.

This process is iterated until all disconnected parts have been joined to the main tree. However, there is no guarantee that the tree is a graph-theoretical tree (acyclic graph). In fact, it almost certainly contains loops where branches have touched each other. In order to estimate branch thickness, it is necessary to cut the loops in the graph (reason covered in method in the next section). A simple heuristic would be to cut loops at the point that is reached contemporaneously from both sides when the entire skeleton graph is traversed starting from the root vertex (e.g., using Dijkstra's algorithm). This turns out to work poorly and cuts the branches in seemingly arbitrary places. A better heuristic is to compute the Euclidean distance from the root node to all vertices on a cycle of the graph and then cut the cycle at the farthest point.

For full details pertaining to both the process of connecting disjoint subtrees and cutting cycles, the interested reader is referred to [23].

2.4.2 Surface Reconstruction

In order to perform a reconstruction of the tree, we need to estimate the diameter of each branch. A simple way to do this is to assign each point of the original point cloud to a segment of the skeleton. Subsequently, we can compute the average perpendicular distance from the sample to the skeleton edge. We can then estimate the radius associated with the skeletal edge as the average of these distances.

Unfortunately, this naive approach leads to a very surprising outcome for some of the smaller branches due to outliers: not every scanned point can be reliably assigned to a single skeletal edge. The issue is illustrated in Fig. 8. Despite only using the 33% of the points closest to the skeleton, outliers still dominate the branch radius estimate for some of the smallest branches.

To resolve this issue, we decided to use an approach where the branch thickness is not directly estimated from the points. We resorted to a principle originally presented by Leonardo da Vinci [13]. da Vinci formulated the hypothesis that when a branch splits into two branches, the cross sectional area of the mother branch is equal to the sum of the cross sectional areas of the daughter branches. In a modern and more flexible formulation [24] expressed the da Vinci tree relationships as

$$d^{\Delta} = \sum_i d_i^{\Delta}, \tag{1}$$

where Δ is the da Vinci exponent, d is the diameter of the mother branch, and d_i are the diameters of the daughter branches. If the da Vinci rule is exactly true as

Fig. 8 The tree reconstructed by simply assigning a cone to each edge. The radius of a given branch segment is estimated from the input point cloud using only the 33% of the points that are closest to that segment of the branch. Still, some estimates are clearly far too big due to outliers

he originally proposed, $\Delta = 2$, but it has also been suggested that this value differs between various types of trees [24], as well as being dependent on location in nature, and amount of destructive influence from humans and animals.

Of course, when a branch splits into several smaller branches, the diameters of the daughter branches need not be identical, but we can express the diameter of the mother branch as a partition of unity,

$$d^{\Delta} = \sum_i d^{\Delta} r_i \ , \tag{2}$$

where $\sum r_i = 1$ and r_i is the proportion of the thickness assigned to branch i. If we plug this into (1), we obtain

$$\sum_i d_i^{\Delta} = \sum_i d^{\Delta} r_i \ . \tag{3}$$

This is clearly true, if

$$d_i = r_i^{1/\Delta} d \ , \tag{4}$$

but we have yet to find r_i. Again, we resort to a heuristic. Since we have a skeleton for the entire tree, we simply compute how many skeletal vertices, $\mathcal{V}_i$, that belong to each daughter branch, i, and $r_i = \mathcal{V}_i / \sum_i \mathcal{V}_i$. r_i now relates directly to the physical size of the daughter branch compared to its siblings, and the branch thickness will be assigned accordingly.

With this approach it is possible to reconstruct the tree using a single constant value for Δ, but results can still be improved. As shown in [23], the approximation is not equally exact near the trunk and for the small twigs. Specifically, we need a bigger Δ value for the small branches in the periphery of the crown. Here, the daughter branches tend to be reconstructed too big compared to the mother branch if the smaller Δ from the thicker branches is used. The variation of Δ across different branch diameters has previously been observed [25]. For this reason, we use a varying Δ value throughout the tree, which depends on the diameter estimated from the original point cloud.

We used the following method to estimate Δ as a function of branch thickness. For each bifurcation, we estimated the thickness of both parent and daughter branches from the point cloud data. From these estimates, we computed the Δ value that best explains the relationship between the diameters of mother and daughter branches at each bifurcation. Finally, we plotted the optimal Δ values against the estimated (mother) branch diameter and fitted a quadratic polynomial to these data points. The result was a quadratic function that maps estimated diameters to Δ values in the range $[2, 4.75]$ which we then proceeded to use for reconstruction of the tree. It is worth pointing out that we still use the original points, but instead of using them to directly estimate the individual branch thicknesses, we use them, in aggregate, to estimate how Δ varies.

Now, everything is in place for a surface reconstruction provided we have an initial estimate of the trunk diameter which is easy to compute from a cross section of the original point cloud near the base of the tree. The da Vinci based algorithm for assigning branch thicknesses throughout the tree then starts from the root node with an initial estimate of d and spreads up the tree, interchanging the role of being a mother and daughter branch automatically.

Having assigned a diameter to each node of the skeleton, we use the method of convolution surfaces [7] to produce an implicit representation of the tree where each branch is represented by a cylindrical surface of correct diameter. The advantage of using convolution surfaces, as opposed to simply representing each edge in the skeleton by a cone stub, is that the implicit primitives blend together forming a coherent surface from which we can construct a polygonal mesh using iso-surface contouring [26].

A rendering of our final reconstruction is shown in Fig. 9, and in Fig. 10 we show how the point cloud differs from the reconstructed surface. The points are color coded according to their distance from the reconstructed mesh. The mean distance is −0.001 m and the standard deviation is 0.025 m. Based on the figure, it is apparent that we are not missing any large branches, but the many small clusters of red points do indicate that the smallest twigs are not captured. It is also clear that parts of the trunk and larger branches are actually inside the reconstructed surface. In part this may be due to imperfect estimations of Δ, but we also observe that the large branches, and especially the trunk, have a far from circular cross section, whereas our reconstruction assumes that the tree is perfectly round.

Fig. 9 In this image, a photograph of the oak tree is compared to a rendering

3 Implementation

The algorithm described above has been implemented using C++. Our implementation is based on the GEL library (https://github.com/janba/GEL) which contains numerous algorithms and data structures for geometry processing. In this paper, we particularly rely on the following data structures and algorithms.

- Creating a graph by connecting points to their closest neighbors requires a data structure for efficiently locating points in space. We employ a kD-tree [27, 28] which is stored space-efficiently as a binary heap. This data structure can be constructed in $O(n \log(n))$ time and provides closest point queries in $O(\log(n))$ time where n is the number of points.
- The graph itself is stored in an adjacency list data structure [29]. Thus, for each node we can query the list of neighbors in constant time.
- The surface of the tree is reconstructed using the method of convolution surfaces. For a point in space, the distance to the convolution surface is simply the distance to the skeleton minus the local radius. We compute a distance field for the convolution surface by visiting each edge of the skeleton and computing the distance to the corresponding part of the convolution surface for each voxel contained in a box associated with the edge. The voxel values are only updated if the distance is smaller than the distance already stored. Effectively, this approach computes the union of convolution surfaces for each branch. Finally, the output mesh is generated using the dual contouring method [26], and the mesh is stored in half-edge data structure [30].

Fig. 10 In the top image, the mesh is shown together with the point cloud. Below only the points are shown. The colors indicate the distance of the points from the mesh: the green points are within a distance of ±4 cm from the surface of the mesh. Red points are further outside and blue are further inside

4 Discussion and Conclusions

Digital geometry is in demand due to its broad utility in application domains that straddle a range of disparate domains from health and life sciences over processing of geographical data to entertainment. While many problems have been effectively solved, and we are now able to deal with enormous geometric data sets, some problems remain.

In this paper, we have considered the situation where the 3D data contains geometric details at multiple scales, and the features on the finest scales are close to the edge of our capabilities for reconstruction. Specifically, very thin tubular structures are hard to reconstruct using traditional methods, but by solving the simpler problem of finding a graph rather than a surface mesh, we are able to ultimately reconstruct a valid surface model from the data.

References

1. Berger, M., Tagliasacchi, A., Seversky, L.M., Alliez, P., Guennebaud, G., Levine, J.A., Sharf, A., Silva, C.T.: A survey of surface reconstruction from point clouds. In: Computer Graphics Forum, vol. 36, pp. 301–329. Wiley Online Library, New York (2017)
2. Kazhdan, M., Hoppe, H.: Screened Poisson surface reconstruction. ACM Trans. Graph. (ToG) **32**(3), 1–13 (2013)
3. Bernardini, F., Mittleman, J., Rushmeier, H., Silva, C., Taubin, G.: The ball-pivoting algorithm for surface reconstruction. IEEE Trans. Vis. Comput. Graph. **5**(4), 349–359 (1999)
4. Digne, J., Morel, J.-M., Souzani, C.-M., Lartigue, C.: Scale space meshing of raw data point sets. In: Computer Graphics Forum, vol. 30, pp. 1630–1642. Wiley Online Library, New York (2011)
5. Siddiqi, K., Pizer, S.: Medial Representations: Mathematics, Algorithms and Applications, vol. 37. Springer, Berlin (2008)
6. Amenta, N., Bern, M., Kamvysselis, M.: A new Voronoi-based surface reconstruction algorithm. In: Proceedings of the 25th Annual Conference on Computer Graphics and Interactive Techniques, pp. 415–421 (1998)
7. Bloomenthal, J., Shoemake, K.: Convolution surfaces. ACM SIGGRAPH. Comput. Graph. **25**(4), 251–256 (1991)
8. Troldborg, N., Sørensen, N.N., Dellwik, E., Hangan, H.: Immersed boundary method applied to flow past a tree skeleton. Agric. For. Meteorol. **308–309**, 108603 (2021). https://doi.org/10.1016/j.agrformet.2021.108603
9. Livny, Y., Yan, F., Olson, M., Chen, B., Zhang, H., El-Sana, J.: Automatic reconstruction of tree skeletal structures from point clouds. ACM Trans. Graph. **29**(6), 1–8 (2010). https://doi.org/10.1145/1882261.1866177
10. Raumonen, P., Kaasalainen, M., Åkerblom, M., Kaasalainen, S., Kaartinen, H., Vastaranta, M., Holopainen, M., Disney, M., Lewis, P.: Fast automatic precision tree models from terrestrial laser scanner data. Remote Sens. **5**(2), 491–520 (2013). https://doi.org/10.3390/rs5020491
11. Hackenberg, J., Morhart, C., Sheppard, J., Spiecker, H., Disney, M.: Highly accurate tree models derived from terrestrial laser scan data: a method description. Forests **5**(5), 1069–1105 (2014). https://doi.org/10.3390/f5051069
12. Xu, H., Gossett, N., Chen, B.: Knowledge and heuristic-based modeling of laser-scanned trees. ACM Trans. Graph. **26**, 19 (2007)

13. Richter, J.P., et al.: The Notebooks of Leonardo da Vinci, vol. 2. Courier Corporation, North Chelmsford (1970)
14. Dellwik, E., van der Laan, M., Angelou, N., Mann, J., Sogachev, A.: Observed and modeled near-wake flow behind a solitary tree. Agric. For. Meteorol. **265**, 78–87 (2019)
15. Angelou, N., Dellwik, E., Mann, J.: Wind load estimation on an open-grown European oak tree. Forestry: An International Journal of Forest Research **92**(4), 381–392 (2019). https://doi.org/10.1093/forestry/cpz026
16. Bærentzen, A., Rotenberg, E.: Skeletonization via local separators. ACM Trans. Graph. **40**(5), 1–18 (2021). https://doi.org/10.1145/3459233
17. Blum, H.: A transformation for extracting new descriptors of shape. Models for the Perception of Speech and Visual Form **19**(5), 362–380 (1967)
18. Cornea, N.D., Silver, D., Min, P.: Curve-skeleton properties, applications, and algorithms. IEEE Trans. Vis. Comput. Graph. **13**(3), 530–548 (2007)
19. Dey, T.K., Sun, J.: Defining and computing curve-skeletons with medial geodesic function. In: Symposium on Geometry Processing, vol. 6, pp. 143–152 (2006)
20. Rebain, D., Angles, B., Valentin, J., Vining, N., Peethambaran, J., Izadi, S., Tagliasacchi, A.: LSMAT least squares medial axis transform. Comput. Graphics Forum **38**(6), 5–18 (2019). Wiley Online Library
21. Tagliasacchi, A., Alhashim, I., Olson, M., Zhang, H.: Mean curvature skeletons. Comput. Graphics Forum **31**(5), 1735–1744 (2012). Wiley Online Library
22. Kordalewski, D.: New greedy heuristics for set cover and set packing. arXiv preprint arXiv:1305.3584 (2013)
23. Villesen, I.B.: Reconstruction of Botanical Trees from Optically Acquired Point Clouds. B.Sc.Eng. Thesis
24. Eloy, C.: Leonardo's rule, self-similarity, and wind-induced stresses in trees. Phys. Rev. Lett. **107**(25), 258101 (2011)
25. Sone, K., Noguchi, K., Terashima, I.: Dependency of branch diameter growth in young acer trees on light availability and shoot elongation. Tree Physiol. **25**(1), 39–48 (2005)
26. Ju, T., Losasso, F., Schaefer, S., Warren, J.: Dual contouring of Hermite data. In: Proceedings of the 29th Annual Conference on Computer Graphics and Interactive Techniques, pp. 339–346 (2002)
27. Bentley, J.L.: Multidimensional binary search trees in database applications. IEEE Trans. Softw. Eng. **SE-5**(4), 333–340 (1979). https://doi.org/10.1109/TSE.1979.234200
28. Berg, M.d., Kreveld, M.v., Overmars, M., Schwarzkopf, O.: Computational geometry. In: Computational Geometry, pp. 1–17. Springer, London, UK (1997)
29. Cormen, T.H., Leiserson, C.E., Rivest, R.L., Stein, C.: Introduction to Algorithms. The MIT Press, Cambridge (2022)
30. Bærentzen, J.A., Gravesen, J., Anton, F., Aanæs, H.: Guide to Computational Geometry Processing: Foundations, Algorithms, and Methods. Springer, London (2012)

Mixed-Integer Programming Models for Two Metal Additive Manufacturing Methods

Jesse Beisegel, Johannes Buhl, Rameez Israr, Johannes Schmidt, Markus Bambach, and Armin Fügenschuh

Abstract Since the beginning of its development in the 1950s, mixed-integer programming (MIP) has been used for a variety of practical application problems, such as sequence optimization. Exact solution techniques for MIPs, most prominently branch-and-cut techniques, have the advantage (compared to heuristics such as genetic algorithms) that they can generate solutions with optimality certificates. The novel process of additive manufacturing opens up a further perspective for their use. With the two common techniques, Wire-Arc Additive Manufacturing and Laser Powder Bed Fusion, the sequence in which a given component geometry must be manufactured can be planned. In particular, the heat transfer within the component must be taken into account here, since excessive temperature gradients can lead to internal stresses and warpage after cooling. In order to integrate the temperature, heat transfer models (heat conduction, heat radiation) are integrated into a sequencing model. This leads to the problem class of MIPDECO: MIPs with partial differential equations as further constraints. We present these model approaches for both manufacturing techniques and carry out test calculations for sample geometries in order to demonstrate the feasibility of the approach.

Keywords Wire-Arc Additive manufacturing · Laser Powder Bed Fusion · Mixed-Integer programming · Partial differential equations · Finite element method · Finite difference method · Optimization

J. Beisegel (✉) · J. Buhl · R. Israr · J. Schmidt · A. Fügenschuh
Brandenburg University of Technology, Cottbus, Germany
e-mail: jesse.beisegel@b-tu.de; johannes.buhl@b-tu.de; rameez.israr@b-tu.de; johannes.schmidt@b-tu.de; fuegenschuh@b-tu.de

M. Bambach
ETH Zurich, Advanced Manufacturing Lab, Zurich, Switzerland
e-mail: mbambach@ethz.ch

E. Cristiani et al. (eds.), *Mathematical Methods for Objects Reconstruction*,
Springer INdAM Series 54, https://doi.org/10.1007/978-981-99-0776-2_5

1 Introduction

In this section we give a brief introduction to the additive manufacturing methods covered in this paper, as well as the simulation of these processes using the finite element method, and mixed-integer programming as our method of choice to model the scheduling of the production process.

1.1 Material Feeding Before the Heating Process

An important group of processes using material feed before heating to the melting point are the powder bed processes. As shown in Fig. 1a, powder is spread over the powder bed via a powder bed feed system. The energy source, for example, a laser beam [45, 59, 64], melts or sinters the powder locally at the points that will later form the component. The excess powder in the build space is not heated and is removed and recycled after the printing process. The layer-by-layer coating and melting, or sintering, process is repeated until the complete geometry is built up. The powder bed system can be used to build up very fine geometries with cavities inside and other high-resolution features. Laser Powder Bed Fusion (LPBF) is selected in this work as an example of trajectory optimization.

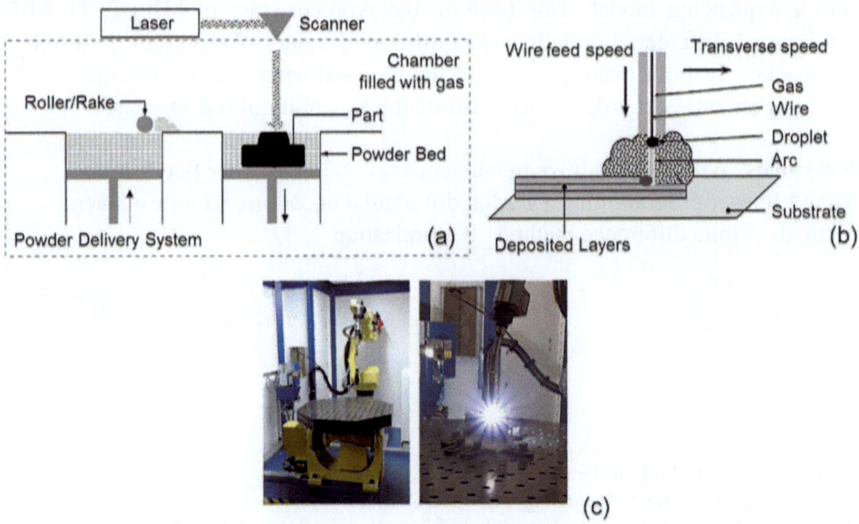

Fig. 1 Principle sketch of the material feed before the heating process (**a**), material feed during the heating process (**b**), and welding robot for deposition on a turn/tilt table (https://www.b-tu.de) (**c**)

1.2 Material Feeding During the Heating Process

In contrast to the LPBF process, in powder feed systems the applied material is delivered in the form of powder into the interaction zone between the heat source and the substrate material. Alternatively, wire can be fed into the heat source or a wire can be used directly as a consumable electrode (see Fig. 1b). When material is fed during the heating process, it is melted with electron beams, laser beams [2], an electric arc, or a plasma [70]. The deposition process is usually performed on a stationary workpiece, with the deposition head moving, or the deposition head stationary and the workpiece moving.

For some years now, a variety of motion systems have been explored, whereby systems such as industrial robots are also used to guide the deposition head and the workpiece is mounted, e.g., on a positioner (see Fig. 1c) [54]. This allows maximum degrees of freedom for positioning the component and the deposition head. Very high deposition rates are possible, especially with arc welding, so that even large-volume components can be built up. A particular challenge of these processes is the trajectory planning of the deposition head, because holes or material accumulation occur if the trajectories and deposition speeds are not optimally set. Material accumulations in nodes very quickly cause process instabilities and generate considerable post-processing costs [54]. With such application systems, even worn or damaged components can be reworked.

1.3 Physical Phenomena During metal WAAM and LPBF Processes

Due to the strong temperature gradients in almost all 3D printing processes, residual stresses build up, which can only be removed with the help of cost-intensive thermal post-treatments. A major disadvantage of AM processes is that, depending on the component geometry and the printing strategy, the intensive heat input also leads to undesirable macroscopic effects in the component. One effect is that a high geometric inaccuracy occurs due to thermal deformations during the printing process and during cooling. In addition, it is possible that components crack during production due to the high thermal stresses [65]. Notch-like defects or plate-stack-like stacking defects can occur on the surface [46, 58] and defects such as gas porosity can occur inside the component [42], which can be controlled by adjusting the process parameters [42, 47]. Macroscopic effects in particular are difficult to avoid and often require additional post-processing such as chemical etching [63] or conventional machining.

These undesirable macroscopic effects mainly depend on the tool path, the process settings, and the cooling strategy. The locally acting heat source causes inhomogeneous thermal expansion, which leads to contraction over the process time and causes mechanical stresses and strains [21]. In almost all AM processes, the temperature profile is strongly dependent on the heat conduction in the component, the heat transfer coefficient, and the radiation into the surrounding medium (powder in LPBF and air in WAAM).

1.4 Application of the FEM Method for Metal WAAM and LPBF Processes

With numerical simulations it is possible to calculate the heat transfer with conduction, convection, and radiation. Thus, in principle, the macroscopic effects such as the distortion, residual stresses, and plastic deformation can be investigated numerically, which has also been done for the WAAM process [9, 19, 38]. Especially when thermal, mechanical, and even the microstructural solvers are used in a coupled way, the simulation time for the LPBF and the WAAM increases tremendously. Due to the complexity of the models, convergence problems occur when solving large deformations, thermal stresses, and residual stresses, which can lead to the premature termination of the computation [7, 11, 20, 48].

Even for larger geometries, the temperature, stress, and thermal distortion could be simulated during the WAAM process and during cooling, with extremely high computation times [55]. With the LPBF method, a simulation of a few hatches is possible [3, 51, 57], but the representation of a layer or even a simple component appears impossible due to the very long laser paths. However, in order to obtain a numerical estimation for real components as well, reference volumes are defined across many layers and calculated with different meso-models [69]. Depending on the resolution of the reference volumes, these approaches enable a very fast global estimation of the temperature and the temperature gradient. Due to the long calculation time, or because the reference volumes are too large, both types of FEM models cannot be used to optimize the trajectory of the heat source with regard to the local temperature gradient in a layer.

1.5 Direct Solution of PDEs for Metal WAAM Processes

An alternative to FEM models is the direct solution of PDEs to calculate the temperature distribution in a layer. For this purpose, a method was first set up to model the nodes of thin-walled hollow structures produced with the WAAM method and to calculate their cooling by thermal radiation. The trajectory of the arc is realized with the sequential activation of the nodes. An optimal node sequence is calculated under the constraints of (i) single node visit, (ii) minimum number of start and end points, and (iii) minimum temperature gradient using MILP [28]. Building on this work, heat conduction was integrated by implementing beam elements (1D) between the nodes [4]. To apply the optimization strategy to the LPBF method, the temperature calculation in the pressure plane (2D) could be adapted, initially considering only the heat conduction in the layer [6]. All works have in common that a steady state process is assumed and the temperature of the lower layer is stable.

1.6 Mixed-Integer Linear Programming

Mixed-Integer Programming refers to the mathematical field of modeling and solving problems from a certain class of optimization problems of the form

$$\begin{aligned} z^* = \min \quad & cx \\ \text{s.t.} \quad & Ax \leq b \\ & x \in \mathbb{Z}^p \times \mathbb{Q}^{n-p}, \end{aligned} \tag{MILP}$$

where c is an n-dimensional row vector, b is an m-dimensional column vector, and A is an m by n matrix, all containing rational numbers. For a fixed integer $p \in \{1, \ldots, n-1\}$ we speak of a *mixed-integer (linear) program*. Here cx is the *objective* and $Ax \leq b$ are the *constraints*. Two special cases are worth mentioning: when $p = 0$ we deal with a *linear program (LP)* and $p = n$ is a pure *integer (linear) program (ILP)*. Any column vector $x \in \mathbb{Z}^p \times \mathbb{Q}^{n-p}$ with $Ax \leq b$ is a *feasible solution* for (MILP). A feasible solution x^* is an *optimal solution* for (MILP) if its objective function value cx^* is equal to z^*.

The study of LPs began in the mid twentieth century with the work of Kantorovich [41], Koopmans [44], and most notably, Dantzig's invention of the Simplex algorithm to solve general LPs [17]—although earlier attempts can be dated back to Fourier [26] and Motzkin [52]. A few years later, LPs with integrality conditions came into focus. Dantzig, Fulkerson, and Johnson suggested integer variables to model binary yes-no decisions, with the Traveling Salesman Problem as prime application example. For the actual solution of such model they introduced the LP relaxation and a cutting plane approach [18].

Since these method are still in use today, let us now give a brief survey on the solution process of (MILP) (for $p > 1$, i.e., MILP or ILP). We first note that these problems are difficult to solve. From a theoretical perspective, they fall into the call of NP-hard problems [29], so that a theoretically efficient algorithm for the solution of general (M)ILP would imply P = NP (which is an open Millennium problem [39]). From a practical perspective, the solution process of MILPs attacks them from two sides, called the primal and the dual side. On the *primal side* one is concerned with finding good feasible solutions fast. This is the realm of heuristics, such as Taboo Search [30], Simulated Annealing [1, 13], or Evolutionary Algorithms [62], to name just a few. When the objective function emphasizes a minimization, as it does in (MILP), then the objective value cx of every primal solution x gives an upper bound on z^*.

On the *dual side* one tries to give lower bounds on z^*. Most prominently, the integrality conditions are dropped (relaxed), so that (MILP) is turned into an LP problem, which can be solved much easier. Searching over a larger space now, an LP feasible solution $\hat{x}$ that satisfies $A\hat{x} \leq b$ gives a lower bound $c\hat{x}$ on z^*. In case $\hat{x} \in \mathbb{Z}^p \times \mathbb{Q}^{n-p}$, this relaxation would give already a feasible solution. However, this rarely happens in practice, where it can be expected that some of the variables with integrality constraints have fractional values in the solution. Then the integrality conditions are gradually re-introduced by either a cutting plane approach, such as Gomory's [32] (see also [14]), or branch-and-bound [16, 49]. In fact, both approaches can be combined to a branch-and-cut method, which was pioneered by Padberg and Rinaldi [56] and Balas et al. [5].

These methods are today readily available in several software packages, such as IBM ILOG CPLEX [36], Gurobi [33], or Fico Xpress [23], which can solve instances of (MILP) in the order of 100,000 variables and constraints. For further details on mixed-integer programming we refer to the textbooks of Nemhauser and Wolsey [53] and Wolsey [68].

In the additive manufacturing application we introduce below, mixed-integer programming is used to formulate the problem of scheduling the production process within one layer. The sequencing of the printing process is modeled by integer (in fact, binary) decision variables, and the constraints represent restrictions on the possible sequences and consequences of routing decisions, such as the material temperature and temperature gradients, which in turn lead to internal stresses and warpage after cooling.

Typical for applied problems with a physical or technical background, they oftentimes come with relations that are not given by explicit functions but by ordinary (ODE) or partial differential equations (PDE) instead. For example, the conduction of heat is described by the heat equation, a well-known parabolic PDE. It is a modeling as well as solving challenge to embed these relations as constraints into the framework of mixed-integer programming. Approaches for the integration of PDEs can be found in Frank et al. [27], where a PDE was integrated into an MILP using a semi-discretization approach. Another example can be found in Buchheim et al. [8] where the authors decompose the problem into an integer linear programming master problem and a subproblem for calculating linear cutting planes. The term MIPDECO for this problem class was suggested by Leyffer [50].

A usual approach to tackle a PDE is to use a finite difference approach for its discretization [61]. If the discretized equation is linear, it can directly be used as a constraint in (MILP). This approach was used, for instance, by Gnegel et al. [31]. However, the downside of this approach is the enormous size of the constraint system (in terms of number of variables and constraints), which is usually beyond the scope even of modern numerical MILP solvers. Furthermore, the structure of discretized PDE systems (such as band matrices) is vastly ignored by current MILP solvers but can be exploited by special-purpose PDE solvers.

2 Wire-Arc Additive Manufacturing

In the process of Wire-Arc Additive Manufacturing, the desired workpiece is built by progressive deposition of weld beads on an underlying substrate using a weld source moving around the working area freely. The wire is molten with high energy from an electrical arc and deposited in droplets to produce weld beads. Although the process does not necessarily require layer-by-layer deposition, we will consider only cases that can be accomplished by slicing the part geometry and depositing material layer-by-layer. For a single layer, the welding trajectory should be continuous since every time the weld source is moved without welding, there is a chance to introduce bonding defects and the quality of the resulting workpiece reduces.

On the other hand, welding parts of the layer more than once lead to material accumulation which affects the shape of the workpiece and increases post-processing efforts. Furthermore, the high temperature of the weld source can cause large temperature gradients with its surrounding, resulting in strain distribution in the welded material, which can lead even to cracks. Thus it is desirable to achieve a homogeneous temperature distribution within the workpiece by adjusting the welding trajectory. Taking these aspects into account, careful planning of the welding trajectory is crucial for process efficiency.

In this work, we consider only workpieces with wall thicknesses larger than the width of the weld bead. Since for areas filled with material there are many possible ways to manufacture them [40], we assume that the path strategy is given and only the sequence of the single welding moves can be optimized. A study about thin-walled structures with wall thickness as broad as the width of the weld bead can be found in [4].

2.1 Path Generation

For the generation of a feasible welding trajectory, we recall the setup from [4] and extend it by incorporating the possibility of multiple non-connected components within one layer and time-dependent transition moves.

We consider a given geometry of a two-dimensional layer as a graph with nodes $i \in \mathcal{V}$ at the coordinates $\mathbf{r}_i \in \mathbb{R}^2$ of every intersection point between two welding segments and edges $(i, j) \in \mathcal{W}$ describing the part of the respective welding segment between nodes $i \in \mathcal{V}$ and $j \in \mathcal{V}$ with length $l_{i,j} \in \mathbb{R}_+$. Let in the following $\mathcal{V}^{odd} \subseteq \mathcal{V}$ and $\mathcal{V}^{even} = \mathcal{V} \setminus \mathcal{V}^{odd}$ denote the set of nodes with odd and even node degree, respectively. In this setting we assume that every welding segment $(i, j) \in \mathcal{W}$ is printed at once and transition moves can only be performed between the nodes. To reduce defects and increase the quality of the workpiece, their number should be minimized.

Since every segment must be welded to process the whole layer, the problem of finding a feasible welding trajectory can be seen as a Chinese postman problem [22] in the graph $G = (\mathcal{V}, \mathcal{W})$. For this problem it is known that, if necessary, additional edges must be inserted between nodes of odd node degree to keep their number minimal. If G is not connected and contains more than one component, this holds for the trajectory within every component and the transition between two components containing nodes of odd node degree. For components without nodes of odd node degree, every node is a possible start- or endpoint for them. Let $\nu \in \mathbb{N}$ denote the number of components of G, $\mathcal{V}_i^{odd}$ and $\mathcal{V}_i^{even}$ the sets of nodes with odd or even node degree in component $i \in \{1, \ldots, \nu\}$, $\mathcal{I} \subseteq \{1, \ldots, \nu\}$ the set of all components with $\mathcal{V}_i^{odd} \neq \emptyset$, and $\mathcal{V}^{tran} = \left(\bigcup_{i \in \mathcal{I}} \mathcal{V}_i^{odd}\right) \cup \left(\bigcup_{i \notin \mathcal{I}} \mathcal{V}_i^{even}\right)$. Thus, all transition moves are also restricted to the set $\mathcal{U} = \mathcal{V}^{tran} \times \mathcal{V}^{tran}$ and the trajectory must start in a node $i \in \mathcal{V}^{tran}$, otherwise another transition move is required. To incorporate transition moves into the model, let $\omega \in \mathbb{N}$ be their minimal number required to process the complete layer and $d_{i,j}^e = \|\mathbf{r}_i - \mathbf{r}_j\|_2$ the Euclidean distance between nodes $(i, j) \in \mathcal{U}$. Note that ω also contains transition moves between components of the graph, if it is not connected.

The maximum velocities of the weld source while welding $v^w \in \mathbb{R}_+$ and for transiting $v^m \in \mathbb{R}_+$ are given parameters of the welding process. Let Δt denote the length of one discrete time step. Thus, the number of time steps to process every welding segment $(i, j) \in \mathcal{W}$ is given by $\tau_{i,j}^w = \left\lceil \frac{l_{i,j}}{v^w \Delta t} \right\rceil$ and the whole layer is processed in $T^{proc} = \sum_{(i,j) \in \mathcal{W}} \tau_{i,j}^w$ time steps. In a similar way, the number of necessary time steps to perform a transition move between nodes $i, j \in \mathcal{V}$ is computed by $\tau_{i,j}^m = \left\lfloor \frac{d_{i,j}^e}{v^m \Delta t} \right\rfloor$. Due to the varying length of the transition moves, the overall time to perform them is not known a priori, but it can be overestimated by $T^{tran} = \omega \max_{i,j \in \mathcal{V}} \tau_{i,j}^m$. Overall, the discrete time horizon is given by $\mathcal{T} = \{1, \ldots, T^{max}\}$ with $T^{max} = T^{proc} + T^{tran}$. As abbreviations we use $\mathcal{T}_0 = \mathcal{T} \cup \{0\}$, $\mathcal{T}^- = \mathcal{T} \setminus \{T^{max}\}$, and $\mathcal{T}^{end} = \{T^{proc} + \omega \min_{i,j \in \mathcal{V}} \tau_{i,j}^m, \ldots, T^{max}\}$, the set of discrete time steps where the process could finish, depending on the length of the necessary transition moves.

Since every segment can be processed in both directions, the edge set $\mathcal{W}$ is expanded to $\overline{\mathcal{W}} = \{(i, j) \in \mathcal{V} \times \mathcal{V} \mid (i, j) \in \mathcal{W} \vee (j, i) \in \mathcal{W}\}$ and the number of time steps for processing is adjusted to $\tau_{i,j}^w = \tau_{j,i}^w$ for $(i, j) \in \overline{\mathcal{W}}$. Relating all possible connections with their respective processing time, we obtain the sets $\mathcal{W}^* = \{(i, t_i, j, t_j) \in \mathcal{V} \times \mathcal{T}_0 \times \mathcal{V} \times \mathcal{T} \mid (i, j) \in \overline{\mathcal{W}}, t_j = t_i + \tau_{i,j}^w\}$ for the welding moves and $\mathcal{U}^* = \{(i, t_i, j, t_j) \in \mathcal{V} \times \mathcal{T} \times \mathcal{V} \times \mathcal{T} \mid (i, j) \in \mathcal{U}, t_j = t_i + \tau_{i,j}^m\}$ for all possible transition moves. Note that transition moves cannot occur in the first time step since if this happens, there is a feasible welding trajectory starting at the end point of this transition with a smaller ω and less processing time.

Binary variables $w_{i,t_i,j,t_j} \in \{0, 1\}$, indicating if the weld source moves from node $i \in \mathcal{V}$ to node $j \in \mathcal{V}$ from time step $t_i \in \mathcal{T}_0$ to time step $t_j \in \mathcal{T}$, are used to track the welding trajectory. For the transition moves, there are binary variables $u_{i,t_i,j,t_j} \in \{0, 1\}$, equal to one if and only if the weld source moves between node $i, j \in \mathcal{V}$ from time step $t_i \in \mathcal{T}$ to time step $t_j \in \mathcal{T}$ without welding. Due to the varying number of time steps for the transition moves, further binary variables

$u_{i,t} \in \{0, 1\}$, indicating whether the welding trajectory ends in node $i \in \mathcal{V}^{tran}$ at time step $t \in \mathcal{T}^{end}$, are required. Thus, a feasible welding trajectory is described by the following constraints: The weld source must start its path at some node

$$\sum_{\substack{i,j,t_j:(i,0,j,t_j)\in\mathcal{W}^* \\ i\in\mathcal{V}^{tran}}} w_{i,0,j,t_j} = 1, \tag{1}$$

$$\sum_{j,t_j:(i,0,j,t_j)\in\mathcal{W}^*} w_{i,0,j,t_j} = 0 \qquad \forall i \in \mathcal{V} \setminus \mathcal{V}^{tran}. \tag{2}$$

Furthermore, every segment must be processed

$$\sum_{t_i,t_j:(i,t_i,j,t_j)\in\mathcal{W}^*} w_{i,t_i,j,t_j} + \sum_{t_j,t_i:(j,t_j,i,t_i)\in\mathcal{W}^*} w_{j,t_j,i,t_i} = 1 \qquad \forall (i,j) \in \mathcal{W}. \tag{3}$$

The resulting trajectory must be continuous

$$\sum_{k,t_k:(k,t_k,i,t)\in\mathcal{W}^*} w_{k,t_k,i,t} + \sum_{k,t_k:(k,t_k,i,t)\in\mathcal{U}^*} u_{k,t_k,i,t} = \sum_{j,t_j:(i,t,j,t_j)\in\mathcal{W}^*} w_{i,t,j,t_j} + \sum_{j,t_j:(i,t,j,t_j)\in\mathcal{U}^*} u_{i,t,j,t_j} \qquad \forall i \in \mathcal{V}^{tran},\ t \in \mathcal{T} \setminus \mathcal{T}^{end}, \tag{4}$$

$$\sum_{k,t_k:(k,t_k,i,t)\in\mathcal{W}^*} w_{k,t_k,i,t} + \sum_{k,t_k:(k,t_k,i,t)\in\mathcal{U}^*} u_{k,t_k,i,t} = \sum_{j,t_j:(i,t,j,t_j)\in\mathcal{W}^*} w_{i,t,j,t_j} + \sum_{j,t_j:(i,t,j,t_j)\in\mathcal{U}^*} u_{i,t,j,t_j} + u_{i,t} \qquad \forall i \in \mathcal{V}^{tran},\ t \in \mathcal{T}^{end}, \tag{5}$$

$$\sum_{k,t_k:(k,t_k,i,t)\in\mathcal{W}^*} w_{k,t_k,i,t} = \sum_{j,t_j:(i,t,j,t_j)\in\mathcal{W}^*} w_{i,t,j,t_j} \qquad \forall i \in \mathcal{V} \setminus \mathcal{V}^{tran},\ t \in \mathcal{T}. \tag{6}$$

Transition moves cannot be used consecutively since they then can be merged to a single one

$$\sum_{k,t_k:(k,t_k,i,t)\in\mathcal{U}^*} u_{k,t_k,i,t} + \sum_{j,t_j:(i,t,j,t_j)\in\mathcal{U}^*} u_{i,t,j,t_j} \le 1 \qquad \forall i \in \mathcal{V}^{tran},\ t \in \mathcal{T}. \tag{7}$$

Finally, the number of end nodes and transition moves is limited by

$$\sum_{i \in \mathcal{V}^{tran}} \sum_{t \in \mathcal{T}^{end}} u_{i,t} = 1, \tag{8}$$

$$\sum_{(i,t_i,j,t_j) \in \mathcal{U}^*} u_{i,t_i,j,t_j} = \omega. \tag{9}$$

2.2 Temperature Calculation

Considering the temperature distribution within the workpiece, it is affected by the heat input of the weld source, heat conduction, convection, and heat radiation. According to [34], a general boundary condition, incorporating convection and radiation and maintaining energy conservation, is given by

$$\frac{\partial \theta}{\partial n}(x, y, t) = -\frac{h}{\lambda}\left(\theta(x, y, t) - \theta^{amb}(x, y, t)\right) - \frac{\varepsilon\sigma}{\lambda}\left(\theta(x, y, t)^4 - \theta^{amb}(x, y, t)^4\right) \quad \forall (x, y) \in \partial\Omega,\ \forall t \in [0, T], \tag{10}$$

with thermal conductivity $\lambda \in \mathbb{R}_+$, heat transfer coefficient $h \in \mathbb{R}_+$, total emissivity $\varepsilon \in \mathbb{R}_+$, Stefan-Boltzmann constant σ, and ambient temperature $\theta^{amb} : \Omega \times [0, T] \rightarrow \mathbb{R}_+$. Following the Rosseland approximation from [4] to linearize the nonlinear radiation term, we achieve a linear approximation of the heat exchange at boundary by

$$\frac{\partial \theta}{\partial n}(x, y, t) = \kappa^e \left(\varphi^{add} - \theta(x, y, t)\right) \qquad \forall (x, y) \in \partial\Omega,\ \forall t \in [0, T]. \tag{11}$$

Its slope $\kappa^e \in (0, 1]$ and additive constant $\varphi^{add} \in \mathbb{R}_+$ must be estimated to achieve appropriate values. Using 11 as boundary condition, the progression of the temperature at every point within the layer can be described by the two-dimensional heat equation

$$\frac{\partial \theta}{\partial t}(x, y, t) = \alpha\left(\frac{\partial^2 \theta}{(\partial x)^2}(x, y, t) + \frac{\partial^2 \theta}{(\partial y)^2}(x, y, t)\right) + q(x, y, t) \quad \forall (x, y) \in \Omega,\ t \in (0, T], \tag{12a}$$

$$\frac{\partial \theta}{\partial n}(x, y, t) = \kappa^e \left(\varphi^{add} - \theta(x, y, t)\right) \qquad \forall (x, y) \in \partial\Omega,\ \forall t \in [0, T], \tag{12b}$$

$$\theta(x, y, 0) = \theta^{init}(x, y) \qquad \forall (x, y) \in \Omega, \tag{12c}$$

with thermal diffusivity $\alpha \in \mathbb{R}_+$ and initial temperature distribution $\theta^{init} : \Omega \to \mathbb{R}_+$. In the following, we assume the heat source function $q(x, y, t) : \Omega \times [0, T] \to \mathbb{R}_+$ to be piece-wise constant.

To transform the partial differential equation system (12) into the discrete framework, we apply the finite element method (FEM) according to [66]. Along every welding segment $(i, j) \in \mathcal{W}$, $\tau^w_{i,j} - 1$ equidistantly distributed discretization points are added and stored in the set $\mathcal{V}^{int}$. Let in the following $\overline{\mathcal{V}} = \mathcal{V} \cup \mathcal{V}^{int}$ denote the set of all nodes, $n = |\overline{\mathcal{V}}|$, and $\xi : \mathcal{V}^{int} \to \mathcal{W} \times \{1, \dots, \tau^w_{i,j}\}$ the function assigning every interior node to its position. Then, the FEM is set up with the node set $\overline{\mathcal{V}}$ as discretization points, the shape of the considered geometry as its boundary, and linear triangle elements between the nodes. Using the implicit time approach, it results in the linear equation system

$$(M + \Delta t K)\boldsymbol{\theta}_{t+1} = \Delta t \left(\mathbf{q}_{t+1}\mathbf{f}^{\mathbf{H}} + \kappa^e \varphi^{add}\mathbf{f}^{\mathbf{R}}\right) + M\boldsymbol{\theta}_t, \tag{13}$$

with mass matrix $M = (m_{i,j}) \in \mathbb{R}^{n \times n}$, stiffness matrix $K = (k_{i,j}) \in \mathbb{R}^{n \times n}$, and load vectors $\mathbf{f}^{\mathbf{H}} \in \mathbb{R}^n$, $\mathbf{f}^{\mathbf{R}} \in \mathbb{R}^n$. Note that the stiffness matrix is computed by $K = \alpha K^S + \kappa^e K^R$, where $K^S \in \mathbb{R}^{n \times n}$ and $K^R \in \mathbb{R}^{n \times n}$ are the effects of (12a) and (12b), respectively. The heat input in all nodes at time step $t \in \mathcal{T}_0$ is given by the vector $\mathbf{q}_t \in \mathbb{R}^n_+$ and the vector $\boldsymbol{\theta}_t$ consists of the variables $\theta_{i,t} \in \mathbb{R}_+$, describing the temperature of node $i \in \mathcal{V}$ at time step $t \in \mathcal{T}_0$. Furthermore, we use f^H_i, f^R_i, and $q_{i,t}$ to denote element $i \in \mathcal{V}$ of the above defined vectors.

For the weld source, we use the piece-wise constant approximation of the Goldak heat source model derived in [4]. It assumes the area of effect of the weld source to be circular with a homogeneous energy distribution in every direction and splits it into K^w non-overlapping rings, where a constant proportion $\kappa^w_1 > \kappa^w_2 > \kappa^w_{K^w}$ of the maximum welding temperature φ^w is added to every node within it. Every ring is identified with the interval $\mathcal{P}_k$, $k = 1, \dots, K^w$, given by the minimum and the maximum distance from the weld source, where the factor κ^w_k applies. Due to the choice of the nodes $\overline{\mathcal{V}}$, the heat source is centered above one node at every time step. Thus, the heat input vector $\mathbf{q}_t$ is given by

$$q_{i,t} = \varphi^w \left(w_{i,t} + \sum_{k=1}^{K^w} \sum_{\substack{j \in \overline{\mathcal{V}} \\ d^e_{i,j} \in \mathcal{P}_k}} \kappa^w_k w_{j,t} \right) \qquad \forall i \in \overline{\mathcal{V}},\ t \in \mathcal{T}_0, \tag{14}$$

where $w_{i,t}$ is an abbreviation for

$$w_{i,t} = \begin{cases} 0 & , i \in \overline{\mathcal{V}} \setminus \mathcal{V}^{odd}, \\ & t = 0, \\ \sum_{j,t_j:(i,0,j,t_j)\in\mathcal{W}^*} w_{i,0,j,t_j} & , i \in \mathcal{V}^{odd},\ t = 0, \\ \sum_{\substack{(h,t_h,j,t_j)\in\mathcal{W}^* \\ \xi(i)=(h,j,k) \\ t=t_h+k}} w_{h,t_h,j,t_j} + \sum_{\substack{(j,t_j,h,t_h)\in\mathcal{W}^* \\ \xi(i)=(h,j,k) \\ t=t_h-k}} w_{j,t_j,h,t_h} & , i \in \mathcal{V}^{int},\ t \in \mathcal{T}, \\ \sum_{h,t_h:(h,t_h,i,t)\in\mathcal{W}^*} w_{h,t_h,i,t} + \sum_{h:(h,t_h,i,t)\in\mathcal{U}^*} u_{h,t_h,i,t} & , i \in \mathcal{V}, t \in \mathcal{T}. \end{cases} \tag{15}$$

Discretizing the initial temperature distribution $\theta^{init}(x, y)$ to $\theta_i^{init} \in \mathbb{R}_+$ describing the initial temperature of node $i \in \overline{\mathcal{V}}$, the temperature distribution within the layer is calculated by

$$\theta_{i,0} = \theta_i^{init} \qquad \forall i \in \overline{\mathcal{V}}, \tag{16}$$

$$\sum_{j\in\overline{\mathcal{V}}} (m_{i,j} + \Delta t k_{i,j})\theta_{j,t} = \sum_{j\in\overline{\mathcal{V}}} m_{i,j}\theta_{j,t-1} + \Delta t\left(q_{i,t} f_i^H + \kappa^e \varphi^{add} f_i^R\right)$$
$$\forall i \in \overline{\mathcal{V}},\ t \in \mathcal{T}. \tag{17}$$

2.3 *Objective Function*

The main difficulty for the construction of the model is the choice of the objective function. Simulating the resulting warpage of a printing process is very complicated, and even in a refined FEM simulation with fixed printing order the warpage can only be inferred by the resultant displacement. One natural way to circumvent this problem is to infer the behavior of the warpage from a feature that is easier to simulate, such as the temperature distribution. Uneven temperature distribution and successive cooling leads to thermal stress in the material, which is a key cause of warpage. Note that temperature distribution is not the only factor that causes warpage. This effect can also be caused by the shape of the object, irregularities in the material and many other factors. This implies that an even temperature distribution does not necessarily imply low warpage; however, it is one important ingredient. From this observation an obvious choice for the optimization objective is the minimization of the occurring temperature gradients by

$$\min \quad \frac{1}{(T^{max}+1)|\mathcal{W}|} \sum_{(i,j)\in\mathcal{W}} \sum_{t\in\mathcal{T}_0} |\theta_{i,t} - \theta_{j,t}|, \tag{18}$$

where we divide by $(T^{max} + 1)|\mathcal{W}|$ to normalize the objective function and return the average thermal gradient within the layer for the computed trajectory. In the following, we refer to this objective as *Grad*.

The objective (18) is complicated since it requires the linearization of many absolute value functions to fit the chosen MILP framework. Furthermore, by assigning the variables $w_{i,t} = \frac{1}{|\overline{\mathcal{V}}|}$ for every node $i \in \overline{\mathcal{V}}$ and time step $t \in \mathcal{T}_0$, the objective value of the linear relaxation can be brought close to zero. This leads to poor bounds within the solution process and can have significant impact on the computational performance, as can be seen at the computations in [4]. In the construction process, the described phenomenon can be seen as a partial welding at all nodes at the same moment, which is not possible in reality.

Another choice for the optimization objective is to minimize the absolute deviation from a given target temperature $\theta_{i,t}^{tar}$ in node $i \in \overline{\mathcal{V}}$ at time step $t \in \mathcal{T}_0$, referred to as *Dev* in the following. With this approach, different goals can be achieved by choosing appropriate values. If the target temperature is chosen constant and equal for all nodes and time steps, a welding trajectory with a homogeneous temperature distribution is preferred. Specified material properties can be obtained by setting the target temperature for the desired nodes to the necessary temperature progression. The normalized objective function for this approach is given by

$$\min \quad \frac{1}{(T^{max} + 1)|\overline{\mathcal{V}}|} \sum_{i \in \overline{\mathcal{V}}} \sum_{t \in \mathcal{T}_0} |\theta_{i,t} - \theta_{i,t}^{tar}|. \tag{19}$$

Since in both objective functions the nonlinear absolute value function is used, additional variables are required to linearize it. Let us denote $\vartheta_{i,j,t}^{+}, \vartheta_{i,j,t}^{-} \in \mathbb{R}_{+}$ the positive and the negative proportion of the absolute value function for every welding segment $(i, j) \in \mathcal{W}$ at time step $t \in \mathcal{T}_0$ in (18) and $\vartheta_{i,t}^{+}, \vartheta_{i,t}^{-} \in \mathbb{R}_{+}$ its respective proportion in node $i \in \overline{\mathcal{V}}$ at time step $t \in \mathcal{T}_0$ in (19). Then, the linear version of objective *Grad* is given by

$$\min \quad \frac{1}{(T^{max} + 1)|\mathcal{W}|} \sum_{(i,j) \in \mathcal{W}} \sum_{t \in \mathcal{T}_0} \left(\vartheta_{i,j,t}^{+} + \vartheta_{i,j,t}^{-}\right), \tag{20}$$

with the additional constraint

$$\theta_{i,t} - \theta_{j,t} = \vartheta_{i,j,t}^{+} - \vartheta_{i,j,t}^{-} \qquad \forall (i, j) \in \mathcal{W},\ t \in \mathcal{T}_0. \tag{21}$$

Similarly, the linearized objective *Dev* is

$$\min \quad \frac{1}{(T^{max} + 1)|\overline{\mathcal{V}}|} \sum_{i \in \overline{\mathcal{V}}} \sum_{t \in \mathcal{T}_0} \left(\vartheta_{i,t}^{+} + \vartheta_{i,t}^{-}\right), \tag{22}$$

with the additional constraint

$$\theta_{i,t} - \theta_{i,t}^{tar} = \vartheta_{i,t}^{+} - \vartheta_{i,t}^{-} \qquad \forall i \in \overline{\mathcal{V}},\ t \in \mathcal{T}_0. \tag{23}$$

2.4 Parameter Estimation

The parameters φ^w, κ^e and φ^{add} are artificial parameters that cannot be identified to physical or material parameters or contain several of them. To achieve good values for them, a parameter estimation is necessary. Therefore, the geometry displayed in Fig. 2 was simulated in LS-DYNA with a predefined welding trajectory for every layer, thermal diffusivity $\alpha = 3.774 \cdot 10^{-6}\frac{\mathrm{m}^2}{\mathrm{s}}$, and time step length $\Delta t = 0.5\mathrm{s}$. Basic information about the AM-modeling technique with death-birth elements in the simulation environment of LS-DYNA can be found in [37]. The geometry with the chosen points used for the calibration is displayed in Fig. 2. To achieve data of the process while in steady state, the temperature of the 10th layer is taken as desired temperature distribution which should be approximated by the mathematical model using a weighted absolute deviation. Let $i \in \mathcal{V}_M$ and $\theta_{i,t}^{sim} \in \mathbb{R}_+$ denote the set of nodes which are identified with the chosen points and their simulated temperature data at time step $t \in \mathbb{T}_0$, respectively. Since the welding trajectory is fixed, only φ^w, κ^e, φ^{add} and the temperature $\theta_{i,t}$ of every node $i \in \mathcal{V}$ at time step $t \in \mathcal{T}_0$ remain as variables. The relevant area of the geometry is modeled using the temperature data of the outer chosen points as boundary set $\mathcal{V}_B$ with a Dirichlet boundary condition. Thus, the optimization model simplifies to

$$\min \sum_{i \in \mathcal{V}_M} \sum_{t \in \mathcal{T}_0} \frac{\theta_{i,t}^{sim}}{M} |\theta_{i,t} - \theta_{i,t}^{sim}|, \tag{24a}$$

$$\theta_{i,0} = \theta_i^{init} \qquad \forall i \in \overline{\mathcal{V}}, \tag{24b}$$

Fig. 2 Geometry used to estimate the parameters of the model using the yellow marked points

$$\sum_{j \in \overline{\mathcal{V}}} (m_{i,j} + \Delta t k_{i,j}) \theta_{j,t} = \sum_{j \in \overline{\mathcal{V}}} m_{i,j} \theta_{j,t-1} + \Delta t \left(q_{i,t} f_i^H + \kappa^e \varphi^{add} f_i^R \right)$$
$$\forall\, i \in \overline{\mathcal{V}} \setminus \mathcal{V}_B,\ t \in \mathcal{T}, \tag{24c}$$

$$\theta_{i,t} = \theta_{i,t}^{sim} \qquad \forall\, i \in \mathcal{V}_B,\ t \in \mathcal{T}, \tag{24d}$$

$$k_{i,j} = \alpha k_{i,j}^R + \kappa^e k_{i,j}^S \qquad \forall\, i, j \in \overline{\mathcal{V}}. \tag{24e}$$

where $M \in \mathbb{R}$ is a constant to scale the weights, $k_{i,j}^R \in \mathbb{R}$ and $k_{i,j}^S \in \mathbb{R}$ are the elements of matrices K^R and K^S, respectively. Due to constraint (24c), model (24) is nonlinear, thus the absolute value function in the objective function (24a) can remain. Its optimal solution was computed using a Trust-region method [12] implemented in the Python package SciPy 1.3.1 [67]. The obtained optimal values for the parameters are $\varphi^w = 1569\,\mathrm{K}$, $\kappa^e = 0.1741$, and $\varphi^{add} = 709.56\,\mathrm{K}$.

2.5 Computational Results

For both objective functions, the resulting MILP consists of constraints (1)–(9), (14)–(17), together with objective function (20) and linearization constraint (21) for objective *Grad* and objective function (22) and linearization constraint (23) for objective *Dev*, respectively. The velocities of the heat source were set to $v^w = 6.66 \cdot 10^{-3} \frac{\mathrm{m}}{\mathrm{s}}$, $v^m = 0.03 \frac{\mathrm{m}}{\mathrm{s}}$, and the time step length is $\Delta t = 1\mathrm{s}$. The thermal diffusivity is again $\alpha = 3.774 \cdot 10^{-6} \frac{\mathrm{m}^2}{\mathrm{s}}$, the results of Sect. 2.4 are applied, and the parameters for the heat source are taken according to [4]. As initial temperature we choose $\theta_i^{init} = 773.15\,\mathrm{K}$ for all nodes $i \in \overline{\mathcal{V}}$ and the target temperature is fixed to $\theta^{tar} = 973.15\,\mathrm{K}$ for all nodes and time steps. They were chosen to achieve a constant temperature within the workpiece over the whole processing time reducing thermal stresses by uniform cooling behavior. To illustrate the advantage of trajectory optimization, we consider again the geometry of Sect. 2.4 and compare the solution of the optimization model to 100 random generated sequences. The geometry contains ten components, nine smaller squares and one surrounding square, displayed in Fig. 3. To reduce the model complexity, we assume that any of the ten squares of the geometry must be processed completely before the next square can be chosen and within a square, the edges are welded counterclockwise starting in the upper left corner. Thus, it remains to find the optimal sequence of the ten squares, leading to 3,628,800 possible trajectories and simulating of all their temperature distributions is ineffective.

Applying the optimization model, implemented in AMPL, the considered instance was solved using IBM ILOG CPLEX 20.1 [36] with a time limit of

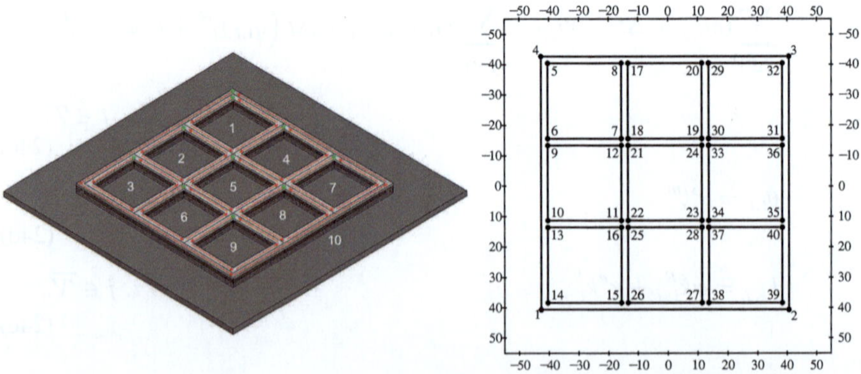

Fig. 3 Example geometry with numbered components and a single layer of it (measurements in mm)

Table 1 Computational results for the considered instance using objectives *Grad* and *Dev*

Objective	*Grad*	*Dev*
Constraints	54,577	87,824
Integer variables	12,301	12,297
Continuous variables	1134	1134
Root node processing time (in sec)	24,890	115,034
Solution time (in sec)	190,921	191,066
Explored nodes	3	6404
Objective function value	–	244.247
Best bound	19.587	236.495
Remaining gap (in %)	100.00	3.28

190,800 s and default settings on a Mac Pro with an Intel Xeon W running 32 threads parallel at 3.2 GHz clock speed and 768 GB RAM. The computational results are recorded in Table 1. Due to the large size of the both models, already the computation of the root node for the branch-and-bound process is time-consuming. The computations time larger than the time limit are caused by the shutdown process requiring also some time. For the objective *Grad* no feasible solution was found during the time limit, indicating the high complexity of this approach. Furthermore, one can see that the best found bound is very low as described in Sect. 2.3. Using the objective *Dev*, the optimization model found a feasible solution with a relatively small gap. It is displayed in Fig. 4. For the random generated sequences, their objective function values were computed after 21,187 s in total using the constraints (14)–(17) with objective function (20) and linearization constraint (21) for objective *Grad* and objective function (22) and linearization constraint (23) for objective *Dev*, respectively. The trajectory related binary variables w_{i,t_i,j,t_j}, $(i, t_i, j, t_j) \in \mathcal{W}^*$, and u_{i,t_i,j,t_j}, $(i, t_i, j, t_j) \in \mathcal{U}^*$, were fixed accordingly. None of the random generated instances is better than the optimized trajectory, as the distribution regarding their objective values in Fig. 5 shows.

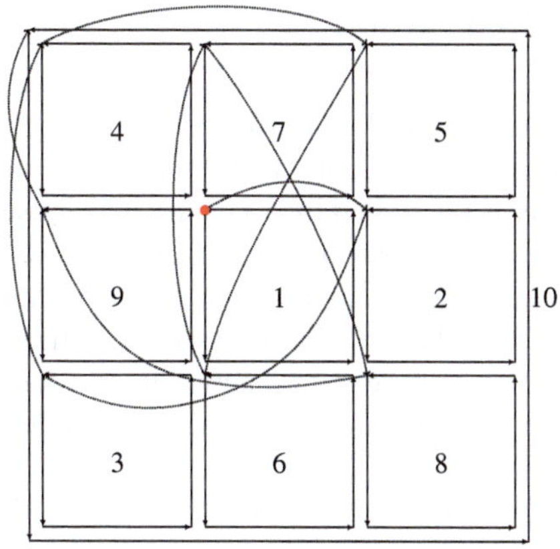

Fig. 4 Best found welding sequence of the considered instance by the optimization model within a time limit of 190,800 s using the objective *Dev*. The red point is the starting node; dashed lines represent transition moves; and the numbers give the sequence of the squares

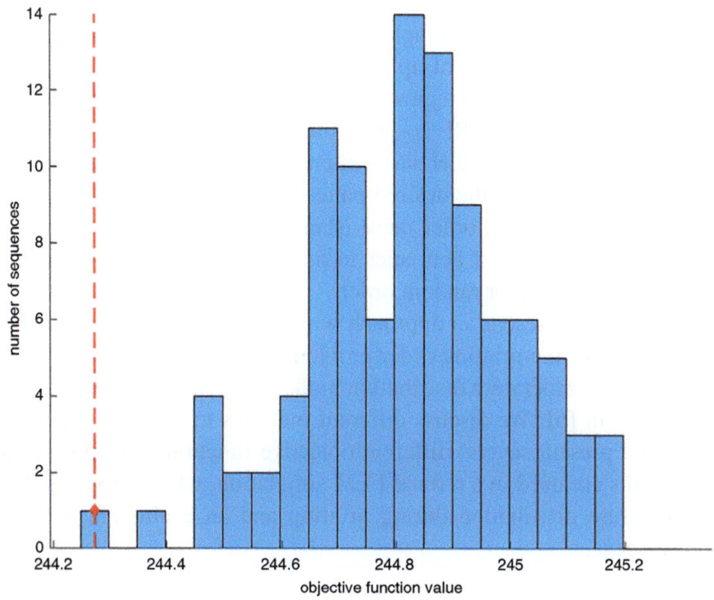

Fig. 5 Distribution of the objective function values of the 100 random generated sequences and the solution found by optimization, marked by a red dotted line

3 Laser Powder Bed Fusion

In laser powder bed fusion (LPBF), a laser beam melts a thin layer of metallic powder, which solidifies during cooling and bonds to the already processed solid material. The laser beam is controlled by a scanner which uses two rotating mirrors to direct the laser to the desired part of the surface. This way the laser can process any point on the surface, and the delay time required to jump between two different areas is much shorter than the actual process time. After finishing one layer, a powder recoater covers the surface with a new thin powder layer, and the melting process iterates, until an entire three-dimensional object emerges. An illustration of this process can be found in Fig. 1.

As the heat source provided by the beam is concentrated on a tiny point of the surface, the temperature of the whole object is usually very uneven. This uneven distribution leads to significant thermal stress and after cooling to warpage. Thermal stress is mainly affected by the material properties, the substrate height and the scanning strategy used by the laser beam [43]. As the material type is usually fixed for a given part and the substrate height is fixed by the machine, the best way to reduce thermal stress and thus warpage is to optimize the scanning strategy. In fact, in [15] the authors show that a simple heuristic on the scanning strategy can lead to a reduction of thermal stress of up to one third compared to a standard strategy. More examples of scanning strategies influencing warpage can be found in [60].

A common strategy for LPBF is called the island strategy. Here, the surface of the printing is divided into smaller surfaces called islands which are then processed consecutively to merge into the desired pattern. In [60] the authors use a heuristic approach to decide in which order these islands are processed. Common to all of these approaches is the use of heuristics to find a good solution, which is in some way better than a standard or random order.

In this section we present an approach which aims to compute mathematically optimal solutions for the printing order problem using a mixed-integer formulation with an integrated temperature distribution model. This model is an extension of the model presented in [6]. We discuss different methods to compute the temperature distribution and present several different objective functions for optimization. The computed results are used in a refined FEM simulation, which gives a good picture of the temperature distribution during printing and an evaluation of the thermal stress and warpage.

3.1 Printing Order

As we are printing with an island strategy, the printing order will describe the sequence in which these islands are processed. Therefore, we will section the surface of the object into equal squares or pixels, each of which describes one island. As the scanner can target any point of the surface at any given time without a significant

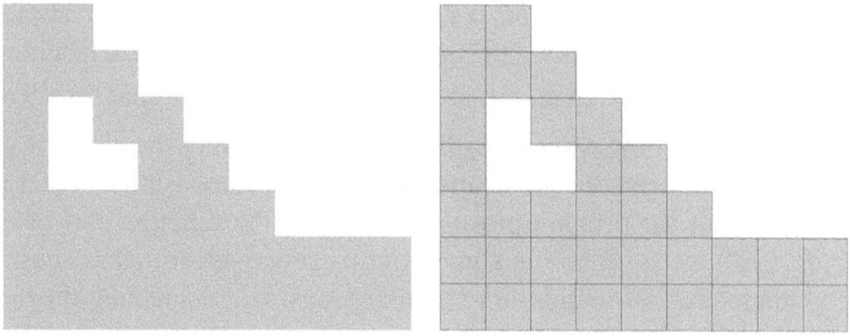

Fig. 6 Example of an object subdivided into pixels

delay, we can assume that any permutation of the islands is an admissible printing order. The path used by the scanner inside the island is usually preset by the given machine and the only information we are given is the printing time of a single island.

For the model we assume that the surface is divided into pixels $(i, j) \in \mathcal{P}$ (see Fig. 6) and the time steps are given in a set $\mathcal{T} = \{1, \ldots, |\mathcal{P}|\}$. The binary decision variable $x_{i,j,t}$ represents the printing decision for pixel (i, j) at time step t, with $x_{i,j,t} = 1$ if (i, j) is printed in time step t and $x_{i,j,t} = 0$ otherwise. We assume that each pixel of the surface must be printed at some time step of the process, which is described by the following equation.

$$\sum_{t \in \mathcal{T}} x_{i,j,t} = 1 \qquad \forall (i, j) \in \mathcal{P}. \tag{25}$$

Due to the production process, we can also assume that at every time step exactly one pixel is printed, which is represented as follows.

$$\sum_{(i,j) \in \mathcal{P}} x_{i,j,t} = 1 \qquad \forall t \in \mathcal{T}. \tag{26}$$

Note that these equations are the classical constraints given for an assignment problem (more on this topic can be found in [10]). With these restrictions it is easy to see that the number of permutations of the pixels $|\mathcal{P}|!$ is equal to the number of admissible printing orders.

3.2 Temperature Calculation

The most common way to mathematically calculate the temperature distribution in a physical object is using Fourier's heat equation:

$$\frac{\partial \theta}{\partial t} - \alpha \nabla^2 \theta = q, \tag{27}$$

where θ is the temperature function, α the thermal diffusivity of the material, and q is a heat source term. Furthermore, the operator ∇^2 describes the Laplacian operator, i.e., $(\frac{\partial^2}{\partial x_1^2} + \ldots + \frac{\partial^2}{\partial x_n^2})$. Given an initial temperature distribution at time 0 and heat sources depending on time and space, the heat distribution of the object is given by θ. In LPBF manufacturing we are dealing with a three-dimensional object in space over the duration of the printing process. Therefore, the function θ is of the form $\theta(x, y, z, t) : \mathbb{R}^4 \to \mathbb{R}$. While in some printing processes it makes sense to deal with the temperature distribution in two dimensions, i.e., neglecting the height (as we will see later for wire-arc additive manufacturing), in this case the build platform acts as a strong heat sink which makes the third dimension essential to modeling. However, a simpler two-dimensional model for LPBF was presented in [6].

Fourier's heat equation is an important example of a *parabolic partial differential equation* and has been widely studied since its introduction the nineteenth century. While there are many approaches to solve these equations analytically, we are interested in a numerical solution which can be embedded into a mixed-integer program. To this end, it is necessary to describe everything with linear (in-) equalities and the procedure should be able to deal with the heat source given by the laser beam. Furthermore, it is not essential that the computation be very precise, as we are mainly interested in the relative distribution of temperature in the object. However, as the overall optimization task will be computationally expensive, the subtask of updating the temperature distribution should be as efficient as possible.

In the case of the heat equation, finite differences produce linear equalities, they can deal with changing momentary heat sources and can be computed quite efficiently. Moreover, this procedure pairs very well with the pixel structure imposed on the printing pattern by the island strategy described in Sect. 3.1. While the computational effort is mostly dependent on the step sizes used in the discretization, it is possible to compromise precision for computational ease here.

For our model, we apply both a common explicit scheme (Forward Time Central Space or FTCS) and an example of an implicit scheme (Backward Time Central Space or BTCS). Being an explicit method, FTCS is computationally very efficient. However, the drawback is that it is only conditionally stable for Fourier's heat equation, i.e., if we use the same step size h for all three dimensions, it is only stable for (see [35]):

$$\Delta t \leq \frac{h^2}{6\alpha}. \tag{28}$$

For the parameters used in our experiments (see Table 2) this condition is far from being fulfilled. The BTCS scheme is an implicit method and unconditionally stable, however, in contrast to the explicit method it is necessary to solve a system of linear equations. While there are several adaptations of BTCS which either improve computation time by generating more structured matrices (ADI) or which have more accurate solutions (Crank-Nicolson), our model computes most efficiently with the standard BTCS scheme. Comparison of these schemes on a test geometry (i.e.,

Table 2 Parameters and values used for computations

Parameter	Unit	Value	Description
Δx	mm	6	Edge length of a pixel
Δz	mm	1.3 and 0.13	Height of a block
Δt	s	3.6864	Time step
λ	$W/(m \cdot K)$	15	Thermal conductivity
ρ	kg/m^3	8000	Density
c	$J/(kg \cdot K)$	500	Specific heat capacity
α_{powder}	m^2/s	$0.03 \cdot \alpha$	Thermal diffusivity unwelded powder
α	m^2/s	3.7510^-6	Thermal diffusivity
τ	W	250	Laser power
β		0.8	Absorptivity
q	K		Heat source
θ^{init}	K	773.15	Initial temperature
θ^{tar}	K	973.15	Target temperature

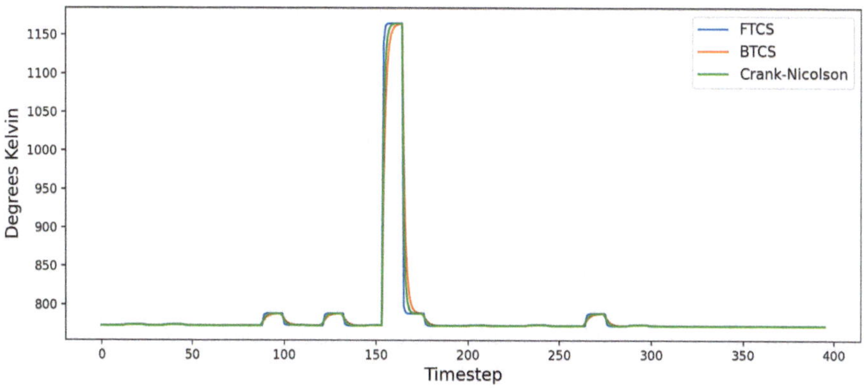

Fig. 7 Here we see a comparison of three finite difference schemes for the temperature development of one block during the simulation of Geometry B (see Fig. 9). In order to implement the FTCS scheme, it was necessary to divide the timestep of printing one pixel by 11. The high temperature peak signifies the point at which the laser welds this particular pixel. The smaller peaks are due to the surrounding pixels being welded

Geometry B, see Fig. 9) shows that the results obtained by FTCS, BTCS, and Crank-Nicolson are very similar in behavior for our model (see Fig. 7).

Following the method of Sect. 3.1 we partition the printing surface into pixels $(i, j) \in \mathcal{P}$ of size $\Delta x \times \Delta x$. The object is then separated into blocks $(i, j, k) \in \mathcal{P}_{obj}$ of size $\Delta x \times \Delta x \times \Delta z$ matching with the pixels of the surface. As the object is surrounded by powder, we embed $\mathcal{P}_{obj}$ in a larger cuboid C which is discretized in the same manner, i.e.,

$$C = \{(i, j, k) : i \in \{1, \dots, N\}, j \in \{0, \dots, M\}, k \in \{0, \dots, H\}\},$$

where N denotes the number of steps in x-direction, M the number of steps in y-direction and H denotes the number of steps in z-direction. All blocks $\mathcal{E} = C \setminus \mathcal{P}$ consist of the metal powder and are modeled with different material qualities. Only the blocks in the bottom-most level, i.e., the set $\{(i, j, k) \in C \ : \ k = 0\}$, are of the same material as the object itself and model the baseplate of the printing machine.

Now we introduce temperature variables $\theta_{i,j,k,t}$ for each block $(i, j, k) \in C$ and each time step $t \in \mathcal{T}$. As the environment of the process is temperature controlled, we are given a fixed initial temperature θ^{init} in time step 0:

$$\theta_{i,j,k,0} = \theta^{init} \qquad \forall (i, j, k) \in C. \tag{29}$$

For the other time steps we use a BTCS discretization scheme applied to the heat equation. In the top layer, i.e., the *printing layer*, the heat transfer is marginal into the surrounding gas. Thus, we assume for simplicity that transfer upwards in z-direction is 0. Furthermore, the temperature equations of the printing layer also include the heat source q which is activated or deactivated by the laser control-variables $x_{i,j,t}$:

$$\begin{aligned}\theta_{i,j,k,t-1} + x_{i,j,t} \cdot q = & \left(1 + 4\alpha \frac{\Delta t}{(\Delta x)^2} + \alpha \frac{\Delta t}{(\Delta z)^2}\right) \theta_{i,j,k,t} \\ & - \alpha \frac{\Delta t}{(\Delta x)^2} \sum_{(i',j') \in N_{i,j}} \theta_{i',j',k,t} - \alpha \frac{\Delta t}{(\Delta z)^2} \theta_{i,j,k-1,t},\end{aligned} \tag{30}$$

where $N_{i,j}$ is the set of blocks adjacent to (i, j) in the printing layer and q is the discretized heat source which will be described in more detail below. For all other layers of the model the heat transfer is computed in all directions and the source q is assumed to be 0, as the laser is only applied to the top layer:

$$\begin{aligned}\theta_{i,j,k,t-1} = & \left(1 + 4\alpha \frac{\Delta t}{(\Delta x)^2} + 2\alpha \frac{\Delta t}{(\Delta z)^2}\right) \theta_{i,j,k,t} \\ & - \alpha \frac{\Delta t}{(\Delta x)^2} \sum_{(i',j') \in N_{i,j}} \theta_{i',j',k,t} - \alpha \frac{\Delta t}{(\Delta z)^2} \left(\theta_{i,j,k+1,t} + \theta_{i,j,k-1,t}\right).\end{aligned} \tag{31}$$

Note that the heat transfer coefficient α here is assumed to be uniform, in order to simplify the equation. In practice it is reliant on the thermal diffusivity of the analyzed material. In this model we are usually given two types of material: the welded solid material and the powder. These two materials have very different thermal diffusivity, which is accounted for in the computations in a later section.

As mentioned above, the environment temperature is controlled throughout the process. The same holds for the baseplate, which is cooled throughout to keep an even temperature. These circumstances suggest the use of *Dirichlet boundary conditions*, i.e., we set the temperature of both the baseplate and the outermost layer of metal powder (see Fig. 8) to the fixed environment temperature throughout.

Fig. 8 The model as seen from above. The pale pixels are part of the object $\mathcal{P}$ and the dark pixels form the surrounding powder

In the process the laser travels in straight lines across the pixel, thereby printing welding beads in the form of stripes. These welding beads are dimensioned to overlap slightly, so as to yield an even distribution of the welded material. In our model this heat source is modeled in a simplified form. We compute the average of this source over the size of the pixel, the depth of the layer and the time taken to print one pixel. This value is used as a constant in the MIP and only turned off or on by the decision variables x, as the energy used by the laser is the same at every timestep and for every pixel. The depth of the layer in our model is usually somewhat larger than the melt depth of the laser.

The basic form of the discretized heat source term q is as follows:

$$q = \frac{S \cdot \Delta t}{\rho \cdot c}, \tag{32}$$

where q is measured in K, S is a volumetric heat source measured in $\mathrm{W/m^3}$, ρ is the density measured in $\mathrm{kg/m^3}$, and c is the specific heat capacity measured in $\mathrm{J/(kg \cdot K)}$. As ρ and c are constants, it remains to compute S.

In our model we compute the heat source term as a pointwise moving heat source which is moving along the x-axis just as the weld beads used in the process. This term can be approximated, due to the fact that the melt depth of 30 µm is much smaller than Δx. Using this value to compute the average over the whole pixel, the volumetric heat source reduces to:

$$S = \frac{\beta P}{2\Delta x \cdot \Delta y \cdot \Delta z}, \tag{33}$$

where β denotes the absorptivity and is assumed to be of value $\beta = 0.8$. For the value q this translates to:

$$q = \frac{\beta P \Delta t}{2\Delta x \cdot \Delta y \cdot \Delta z \cdot \rho c}. \tag{34}$$

3.3 *Objective Functions*

The optimization objective Sect. 2.3 also applies here, thus we can derive possible objective functions in a similar way using the notation of this section. For the minimization of thermal gradients between two blocks, again referred to as objective *Grad*, incorporating additional variables $\vartheta^+_{i,j,i',j',t}, \vartheta^-_{i,j,i',j',t} \in \mathbb{R}_+$ leads to the linearized objective

$$\min \frac{1}{|\mathcal{T}||\mathcal{P}|} \sum_{t\in\mathcal{T}} \sum_{(i,j)\in P} \sum_{(i',j')\in\mathcal{N}_{i,j}} \left(\vartheta^+_{i,j,i',j',t} + \vartheta^-_{i,j,i',j',t}\right), \tag{35}$$

where we divide by $|\mathcal{T}| \cdot |\mathcal{P}|$ to normalize by the number of time steps and pixels in the geometry and the additional constraint

$$\vartheta^+_{i,j,i',j',t} - \vartheta^-_{i,j,i',j',t} = \frac{\theta_{i,j,\text{top},t} - \theta_{i',j',\text{top},t}}{\Delta x} \quad \forall (i,j) \in \mathcal{P}, (i',j') \in \mathcal{N}_{i,j}, \; t \in \mathcal{T}. \tag{36}$$

Here we sum up only the gradients of the printing layer for the sake of simplicity. However, it is also possible to sum up over all blocks of the object.

This objective is not only complicated and involves the computation of many absolute values but one can also easily see by assigning the value $\frac{1}{|\mathcal{P}|}$ to each variable $x_{i,j,t}$ that the value of an LP relaxation is very close to zero. This can be interpreted as spreading the energy of the laser beam evenly across the whole surface of the object. In fact, in practice this is a significant problem and, for example, a test using the geometry seen in Fig. 8 could not be solved with a gap exceeding 95%, even if run for multiple days. In [6] the authors use the same objective function for a much simpler two-dimensional model and have similar difficulties.

These circumstances show that different objectives are needed in order to solve this problem efficiently. As in Sect. 2.3, another possibility is to evaluate the deviation in temperature from a given fixed target temperature θ^{tar} and sum these up for all blocks in the printing layer and all time steps:

$$\min \frac{1}{|\mathcal{T}||\mathcal{P}|} \sum_{t\in T} \sum_{(i,j)\in\mathcal{P}} \left|\theta_{i,j,\text{top},t} - \theta^{tar}\right|, \tag{37}$$

where we again divide by $|\mathcal{T}| \cdot |\mathcal{P}|$ to normalize by the number of time steps and pixels in the geometry and the absolute value function is linearized just as in the objective *Grad*.

As can be seen in the computational results presented in the next section, this already significantly improves running times and yields good results for the printing order. In the following we will refer to this objective as *Dev*. However, in order to make the computations more efficient we also present the following objective:

$$\min \frac{1}{|\mathcal{T}||\mathcal{P}|} \sum_{t \in T} \sum_{(i,j) \in \mathcal{P}} \theta_{i,j,\text{top},t}. \tag{38}$$

In this function we sum up all temperatures in the printing layer for all blocks and all time steps. In the following we will refer to this objective as *Sum*. While at first glance the value computed here seems trivial, i.e., minimizing the amount of heat added to the surface layer by the laser beam, this can also be interpreted as maximizing the heat passed to the boundary in form of the powder bed and the baseplate. As the printing layer will always be the hottest part of the object, it creates a large amount of the thermal stress within the object. Therefore, choosing a printing order which keeps this area as cool as possible can make an impact on the resulting warpage. This is the case even more so if the object is very irregular in the z-direction, for example, in the form of an overhang. In this case, the material on the overhang cannot pass heat toward the baseplate as efficiently as material which has a direct connection to the baseplate. However, for objects that are very regular in the z direction, the conduction we will see in the following section that this objective yields much faster computation times than both of the other presented objectives. While we cannot give a definitive reason for this difference in computational effort, one can note that all terms in this objective function are positive integers, as opposed to absolute values of differences. This makes the problem more closely related to the *generalized assignment problem* which, while still $\mathcal{NP}$-hard [24], can be approximated up to a guarantee of $1 - \frac{1}{\epsilon}$ (see [25]).

3.4 Computational Results

In order to test our model we used a standard setup for the printing process. The material chosen was stainless steel of type 1.4571 whose material properties can be found in Table 2. The shapes to be printed were designed to be asymmetrical, with some thinner areas and a hole. Both instances can be found in Fig. 9 with their respective surface and pixel structure. More irregular shapes are more likely to warrant a customized printing order and edges and holes are likely to cause more thermal stress. We use an edge length for each pixel of 6 mm. With regard to the height, we tested two different approaches. The height of the printed object is chosen as 1.3 mm.

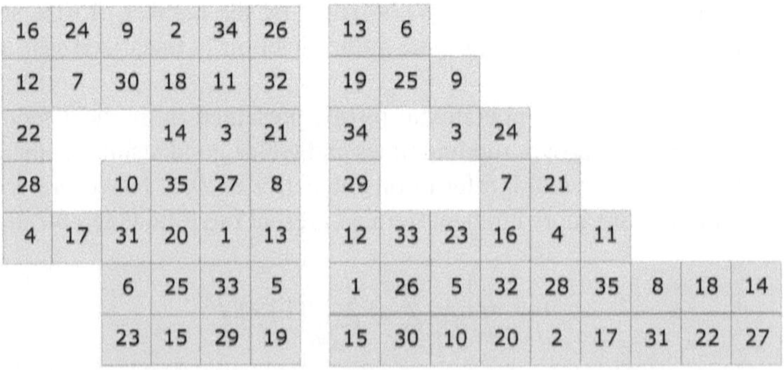

Fig. 9 The image on the left represents the surface of Geometry A, and on the right we see Geometry B. The strategy *big-dev* is given by the numbering on the pixels

As an actually printed layer has a height of approximately 30 μm, we would have to model about 45 layers to reach a height of 1.3 mm. Simulating a thinner object is not as interesting, as the temperature needs several rounds of printing to accumulate in the object. Therefore, we first use one height layer (plus the boundary condition representing the baseplate) and a heat source which is adjusted to this height. This approximates the process in which all printing layers use the same order. In a second approach, we use ten height layers of 0.13 mm each and optimize the strategy for each of these subsequently. This approximates the process where about 4–5 subsequent layers use the same strategy and then a new strategy is computed using the previous temperature distribution. We assume a laser beam of 250 W and as a starting temperature we choose 773.15 K. The heat source is computed using the laser energy and is adjusted for the time step and pixel size.

These setups were then computed on the two different geometries (*Geometry A* and *Geometry B* in Fig. 9) for all presented objectives. The strategies denoted by *big* correspond to the setup using one layer of 1.3 mm. The strategies denoted by *layer* correspond to the setup where ten layers of 0.13 mm were used and are numbered for the layers processed. The different objectives are denoted by $grad$, dev, and sum. The computations were made using IBM ILOG CPLEX 20.1.0.0 [36] with a time limit of 3600 s for the *layer-dev* and *layer-sum* strategies and a time limit of 36 000 s for the *big-dev* and *big-sum* strategies. The LP-method was set to *parallel* and the *numerical emphasis* was switched on to improve the performance of LP relaxation. In the case of the *layer-dev* and *layer-sum* strategies, we used MIP-starts in CPLEX to pass the solution of a previous layer to the next computation. All computations were performed on a Linux system with a Intel Xeon Gold 6136 CPU at 3 GHz using up to 32 kernels and 240 GiB RAM.

The computational results in Table 3 for the *big* model clearly show that the objective *Grad* leads to very inaccurate results when executed in the fixed time span of 36,000 s, while the optimality gap for *Dev* is quite low in the same time frame. Objective *Sum* on the other hand computes to optimality in less than 2 s. Also, in Table 4 one can see that the model is smallest for *Sum*, followed by *Dev* and

Table 3 Computational results displaying runtime, optimality gap, and objective value for all constructed models. The objective values denote the average temperature of the topmost blocks for *Sum*, the average deviation from θ^{tar} for *Dev* and the average temperature gradient for *Grad*. Note that *layer10-dev* did not compute a solution in the defined time frame of 3600*s*

	Geometry A			Geometry B		
Strategy	Runtime	Gap	Value in K	Runtime in s	Gap	Value in K
big-grad	36,000	95.50%	121,184.19	3600	94.47%	114,286.61
big-dev	36,000	1.20%	391.58	36,000	1.45%	393.46
layer1-dev	2.12	0%	197.29	3.72	0%	197.29
layer2-dev	3.85	0%	194.58	4.99	0%	194.58
layer3-dev	3602	2.66%	197.12	3601	2.66%	197.13
layer4-dev	3600	5.23%	199.61	3601	5.23%	199.62
layer5-dev	3603	7.67%	201.97	3611	7.68%	201.99
layer6-dev	3603	10.00%	204.2	3601	10%	204.22
layer7-dev	3604	12.2%	206.26	3620	12.21%	206.30
layer8-dev	3604	14.29%	208.15	3601.73	14.3%	208.2
layer9-dev	3607	16.26%	209.85	3627.77	16.27%	209.92
layer10-dev	3602	18.13%	211.38	3600	–	–
big-sum	1.43	0%	1039.9	1.49	0%	1038.78
layer1-sum	0.41	0%	775.86	0.55	0%	775.86
layer2-sum	3.31	0%	778.57	5.47	0%	778.57
layer3-sum	7.73	0%	781.27	15.31	0%	781.27
layer4-sum	82.35	0%	783.98	125.82	0%	783.97
layer5-sum	130.04	0%	786.67	81.46	0%	786.67
layer6-sum	52.65	0%	789.37	165.17	0%	789.36
layer7-sum	51.05	0%	792.06	113.87	0%	792.04
layer8-sum	141.87	0%	794.74	347.90	0%	794.72
layer9-sum	99.68	0%	797.42	448.05	0%	797.40
layer10-sum	709.81	0%	800.09	360.69	0%	800.07

Table 4 Number of variables and constraints for Geometry B with objectives *Grad*, *Dev*, and *Sum* for the setup using one layer

Objective	*Grad*	*Dev*	*Sum*
Constraints	5047	4492	3197
Binary variables	1225	1225	1225
Continuous variables	6827	5675	3127

then *Grad*. In the *layer* model we can see the effect of the inclusion of multiple height layers. Note that the use of solutions when moving from layer_i to layer_{i+1} leads to some jumps in running time, as can be seen for *layer5-sum* and *layer6-sum*. However, apart from this the computation time jumps in an order of magnitude every other layer. This is an issue when computing with a large amount of layers and further motivates, why we did not use the printing height of 30 μm.

To validate the results of the optimization, the computed printing orders were used in a refined FEM simulation which can generate the temperature distribution, as well as material displacement during the printing process. Here, we present the

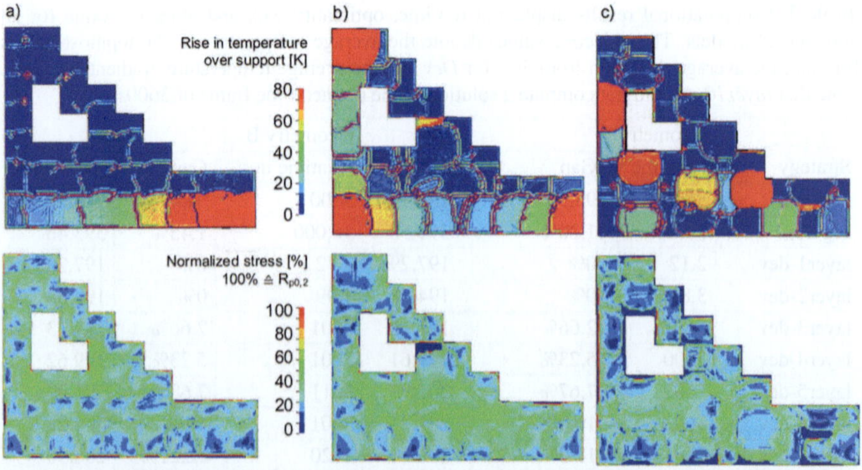

Fig. 10 A visualization of the temperature distribution and the normalized stress in the last step of the process using a standard strategy, as well as the optimized *Sum* and *Dev* strategies on Geometry B. (**a**) Standard. (**b**) Sum. (**c**) Dev

results for the orders generated by the *big-dev* and *big-sum* computations for both presented geometries. A visualization of the printing orders for *Dev* can be found in Fig. 9. Each of these results is compared to a standard printing order which prints the objects in stripes moving from left to right and from top to bottom. In the following, the temperature distribution directly after the last exposure step is shown. Since the heat is distributed very quickly, the scale is chosen in such a way that only a temperature increase up to 80 K due to the heat source is resolved.

For Geometry B the temperature distribution and the normalized stress for the optimized and the standard strategies are given in Fig. 10. We see that for Geometry B the temperature distribution is more uniform for the optimized strategies, while with the standard strategy the temperatures are much higher in the lower half of the object. The stress normalized to the yield stress is slightly reduced for the *Dev* printing sequence.

The displacement produced by the temperature distribution (see Fig. 11) shows clear differences between the strategies. First, the local z-shift is particularly visible in the last processed pixels, which can be seen as red dots. With the standard strategy, the local z-shift can be seen in most areas of the surface. The highest relative maximal resultant displacement of 100% occurs in the *Sum* strategy. This resultant displacement is essentially an aggregate of the displacement in all three dimensions and is only about 60–75% for the *Dev* strategy in comparison to the standard strategy and the *Sum* strategy. Furthermore, it is much better distributed in the *Dev* strategy. For the standard strategy, the maximum value is in the lower right corner, which is the last area to be processed by this strategy.

For Geometry A we see a similar picture with regard to the temperature distribution (see Fig. 12). Due to the processing in stripes of the standard strategy,

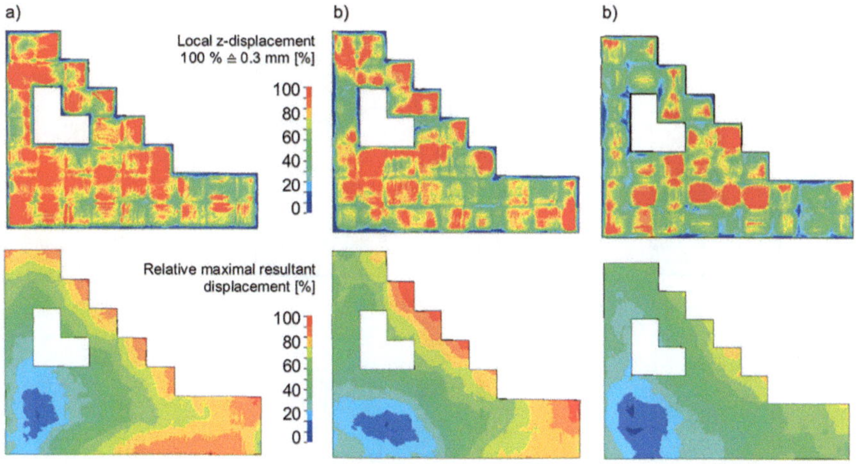

Fig. 11 A visualization of the z-displacement and the relative maximal resultant displacement in the last step of the process using a standard strategy, as well as the optimized *Sum* and *Dev* strategies on Geometry B. (**a**) Standard. (**b**) Sum. (**c**) Dev

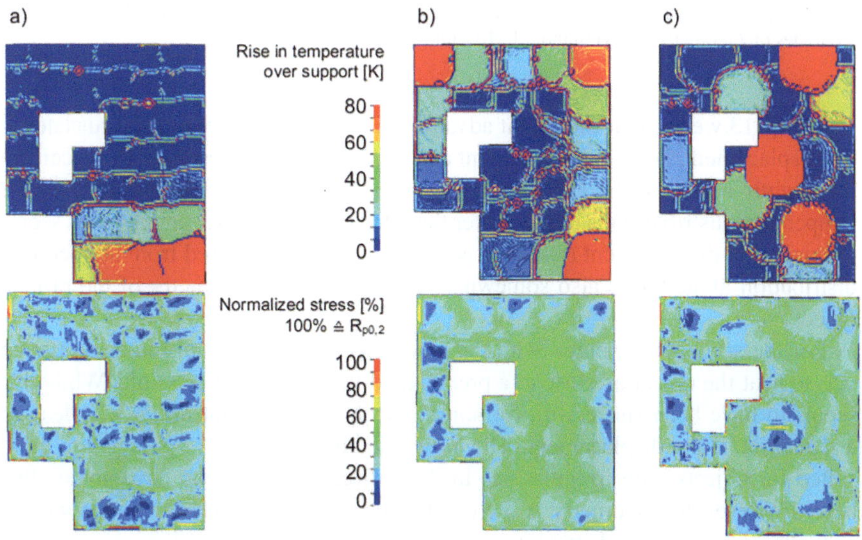

Fig. 12 A visualization of the temperature distribution and the normalized stress in the last step of the process using a standard strategy, as well as the optimized *Sum* and *Dev* strategies on Geometry A. (**a**) Standard. (**b**) Sum. (**c**) Dev

the lower half has much higher temperatures than the top, dividing the object into two temperature zones, while the optimized strategies are again more even. When analyzing the normalized stress, however, the standard strategy seems to have a slight advantage.

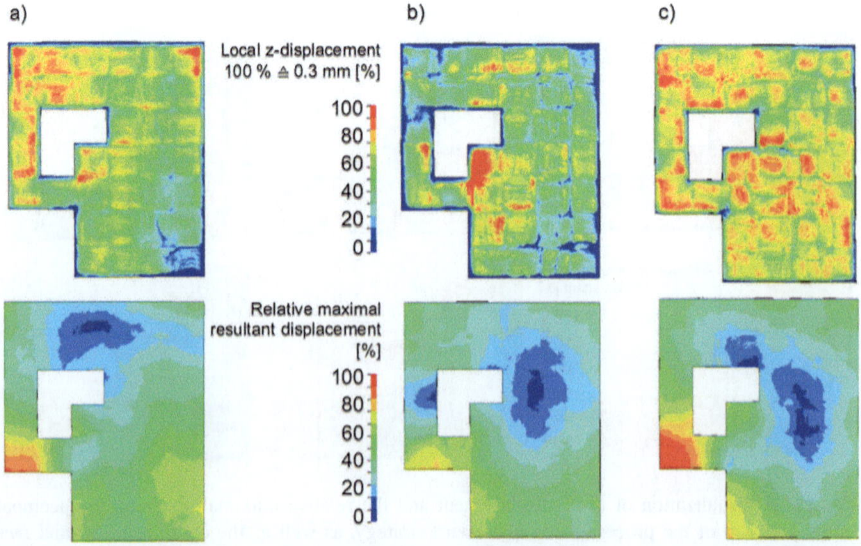

Fig. 13 A visualization of the z-displacement and the relative maximal resultant displacement in the last step of the process using a standard strategy, as well as the optimized *Sum* and *Dev* strategies on Geometry A. (**a**) Standard. (**b**) Sum. (**c**) Dev

In Fig. 13 we see that this slight advantage in normalized stress also translates to the displacement. Both z-displacement and relative maximal resultant displacement are slightly higher in the *Dev* strategy, especially in the corner on the left. As the temperature distribution is in fact better than that of the standard strategy, this seems to imply that displacement (and thus warpage) cannot be inferred from temperature distribution alone but is also somewhat dependent on the geometry of the printed object.

For the strategy computed with the *Dev*-objective, the goal of the computation was to heat the object as close to a preassigned temperature as possible. While this seems to have been successful, for some objects this does not necessarily lead to a reduction of the displacement. The *Sum* objective function is by its nature more suited for objects with irregularities in the z-direction which we have not studied here. However, for Geometry A the computed strategy still performs quite well in comparison to a standard strategy.

Summing up, these results are somewhat inconclusive, as the optimized strategies, while performing well in some points, do not necessarily result in less displacement for all the studied examples. However, the FEM simulation has a model for the laser movement as a heated line. Therefore, the FEM simulation takes into account the direction of the laser beam within a hatch field. The laser movement and the direction of the beam are not available in the MIP model and have been randomly chosen in the FEM simulation. On the other hand, the modeling of the simulated experiments is highly simplified. A 1.33 mm thick plate, for example, was created and is not mechanically fixed. This means that the plate can expand freely, which is, of course, not the case in real 3D printing. In addition, high-resolution simulated experiments with many nodes require a prohibitive computational capacity. We chose 114,752 nodes for the component and 370 nodes for the support to reduce computation time.

Lastly, the differences in displacement are very small (for example, the highest z-displacement is 0.3 mm over all examples). Due to restrictions in computation time, we have only computed small and very thin examples. When more layers are printed and the temperature has time to accumulate, one can expect larger displacement and the printing order to have a higher impact on the overall outcome of the printing process.

4 Conclusions and Future Work

In this work we presented a mixed linear integer programming approach toward the optimization of printing orders for WAAM an LPBD manufacturing. We constructed complete models using real-world parameters, which were tested for computational properties for several instances and objectives. The computed results were evaluated in part using simulation tools common in the field of additive manufacturing, showing promising results for the reduction of warpage, due to the production process.

In future work, the aim is to accelerate the solution process of the WAAM optimization model to reduce the computation time for finding optimal welding trajectories by examining different objective functions and solution methods. Furthermore, the model could be extended to geometries with arbitrary wall strengths and the objective function is reworked to achieve results easier to interpret.

For the optimization of LPBD process many questions are still open. Our model can still be refined with regards to the temperature computation. Other numerical methods to solve the heat can still be tested with regards to the trade-off between accuracy and speed of computation in the integrated MILP model. Furthermore, as the direct simulation of the warpage is already computationally very expensive (the model used here needs several hours to compute one printing layer), the question still remains whether optimizing the heat distribution is the only way to approximate warpage efficiently. Our model can be adapted for a multitude of different objective functions, which could be compared to the ones used here. In the next step, experimental results can be used to further reinforce our results. This seems to be a challenge, as small differences during the printing process and in the material can sometimes lead to different warpage even for the same strategy.

Another question is whether our model can be adapted to other related printing processes. For example, LPBF can also be executed with multiple lasers, both for welding and for heating the material under the melting point. This would lead to very different printing patterns and could have interesting effects on the optimization process.

Acknowledgments This work is part of the project "MALEDIF: Maschinelles Lernen für die additive Fertigung" and was funded by the European Regional Development Fund (ERDF) within the program StaF (Stärkung der technologischen und anwendungsnahen Forschung an Wissenschaftseinrichtungen) under the grant number 85037495.

The authors would like to thank Katharina Eissing, Qui Lam Nguyen, and Felix Jensch for their valuable input and fruitful discussions toward these results.

References

1. Abramson, D., Randall, M.: A simulated annealing code for general integer linear programs. Ann. Oper. Res. **86**, 3–21 (1999)
2. Ansari, M., Jabari, E., Toyserkani, E.: Opportunities and challenges in additive manufacturing of functionally graded metallic materials via powder-fed laser directed energy deposition: a review. J. Mater. Process. Technol. **294**, 117117 (2021)
3. Ansari, P., Salamci, M.U.: On the selective laser melting based additive manufacturing of AlSi10mg: the process parameter investigation through multiphysics simulation and experimental validation. J. Alloys Compd. **890**, 161873 (2022)
4. Bähr, M., Buhl, J., Radow, G., Schmidt, J., Bambach, M., Breuß, M., Fügenschuh, A.: Stable honeycomb structures and temperature based trajectory optimization for wire-arc additive manufacturing. Optim. Eng. **22**(2), 913–974 (2021)
5. Balas, E., Ceria, S., Cornuéjols, G., Natrjay, N.: Gomory cuts revisited. Oper. Res. Lett. **19**, 1–9 (1996)
6. Bambach, M., Fügenschuh, A., Buhl, J., Jensch, F., Schmidt, J.: Mathematical modeling and optimization for powder-based additive manufacturing. Procedia Manufacturing **47**, 1159–1163 (2020)
7. Bellet, M., Hamide, M.: Direct modeling of material deposit and identification of energy transfer in gas metal arc welding. Int. J. Numer. Methods Heat Fluid Flow **23**(8), 1340–1355 (2013)
8. Buchheim, C., Kuhlmann, R., Meyer, C.: Combinatorial optimal control of semilinear elliptic PDEs. Comput. Optim. Appl. **70**(3), 641–675 (2018)
9. Buhl, J., Israr, R., Bambach, M.: Modeling and convergence analysis of directed energy deposition simulations with hybrid implicit/explicit and implicit solutions. Journal of Machine Engineering **19**(3), 94–107 (2019)
10. Burkard, R., Dell'Amico, M., Martello, S.: Assignment Problems: Revised Reprint. SIAM, New York (2012)
11. Chiumenti, M., Cervera, M., Salmi, A., de Saracibar, C.A., Dialami, N., Matsui, K.: Finite element modeling of multi-pass welding and shaped metal deposition processes. Comput. Methods Appl. Mech. Eng. **199**(37–40), 2343–2359 (2010)
12. Conn, A.R., Gould, N.I., Toint, P.L.: Trust Region Methods. SIAM, New York (2000)
13. Connolly, D.: General purpose simulated annealing. J. Oper. Res. Soc. **43**, 495–505 (1992)
14. Cornuéjols, G.: Revival of the Gomory Cuts in the 1990s. Ann. Oper. Res. **149**, 63–66 (2007)
15. Dai, K., Shaw, L.: Distortion minimization of laser-processed components through control of laser scanning patterns. Rapid Prototyp. J. **8**(5), 270–276 (2002)
16. Dakin, R.: A tree-search algorithm for mixed integer programming problems. Comput. J. **8**(3), 250–255 (1965)
17. Dantzig, G.B.: Linear programming. In: Problems for the Numerical Analysis of the Future, Proceedings of the Symposium on Modern Calculating Machinery and Numerical Methods, UCLA (July 29–31, 1948). Applications of Mathematics, vol. 15, pp. 18–21. National Bureau of Standards, Gaithersburg (1951)
18. Dantzig, G., Fulkerson, R., Johnson, S.: Solution of a large-scale traveling-salesman problem. J. Oper. Res. Soc. Am. **2**(4), 393–410 (1954)

19. Das, R., Cleary, P.: Three-dimensional modelling of coupled flow dynamics, heat transfer and residual stress generation in arc welding processes using the mesh-free SPH method. J. Comput. Sci. **16**, 200–216 (2016)
20. Ding, J., Colegrove, P., Mehnen, J., Ganguly, S., Almeida, P.S., Wang, F., Williams, S.: Thermo-mechanical analysis of wire and arc additive layer manufacturing process on large multi-layer parts. Comput. Mater. Sci. **50**(12), 3315–3322 (2011)
21. Ding, D., Pan, Z., Cuiuri, D., Li, H.: Wire-feed additive manufacturing of metal components: technologies, developments and future interests. Int. J. Adv. Manuf. Technol. **81**(1–4), 465–481 (2015)
22. Edmonds, J., Johnson, E.L.: Matching, Euler tours and the Chinese postman. Math. Program. **5**, 88–124 (1973)
23. Fair Isaac Corporation: FICO Xpress Optimizer Reference Manual (2021). https://www.fico.com/fico-xpress-optimization/docs/
24. Fisher, M.L., Jaikumar, R., Van Wassenhove, L.N.: A multiplier adjustment method for the generalized assignment problem. Manag. Sci. **32**(9), 1095–1103 (1986)
25. Fleischer, L., Goemans, M.X., Mirrokni, V.S., Sviridenko, M.: Tight approximation algorithms for maximum general assignment problems. In: Proceedings of the Seventeenth Annual ACM-SIAM Symposium on Discrete Algorithm, pp. 611–620 (2006)
26. Fourier, J.: Solution d'une question particuliere du calcul des inegalities. In: Oeuvres de Fourier, pp. 317–319. G. Olms, Hildesheim (1970). Reprinted of the original 1826 paper with an abstract of an 1824 paper.
27. Frank, M., Fügenschuh, A., Herty, M., Schewe, L.: The coolest path problem. Networks Heterogen. Media **5**(1), 143–162 (2010). https://doi.org/10.3934/nhm.2010.5.143
28. Fügenschuh, A., Bambach, M., Buhl, J.: Trajectory optimization for wire-arc additive manufacturing. In: Operations Research Proceedings, pp. 331–337. Springer International Publishing, Berlin (2019)
29. Garey, M., Johnson, D.: Computers and Intractability: A Guide to the Theory of NP-Completeness. W.H, Freeman and Co, New York (1979)
30. Glover, F.: Future paths for integer programming and links to artificial intelligence. Comput. Oper. Res. **13**(5), 533–549 (1986)
31. Gnegel, F., Fügenschuh, A., Hagel, M., Leyffer, S., Stiemer, M.: A solution framework for linear PDE-constrained mixed-integer problems. Math. Program. Series B **188**, 695–728 (2021)
32. Gomory, R.: An algorithm for the mixed integer problem. Tech. rep., RM-2597, The RAND Cooperation, California (1960)
33. Gurobi Optimization, LLC: Gurobi Optimizer Reference Manual (2021). https://www.gurobi.com
34. Hahn, D.W., Özisik, M.N.: Heat conduction. Wiley, New York (2012)
35. Hindmarsh, A., Gresho, P., Griffiths, D.: The stability of explicit Euler time-integration for certain finite difference approximations of the multi-dimensional advection-diffusion equation. Int. J. Numer. Methods Fluids **4**(9), 853–897 (1984)
36. International Business Machines Corporation: IBM ILOG CPLEX Optimizer Reference Manual (2021). https://www.ibm.com/docs/en/icos/20.1.0
37. Israr, R., Buhl, J., Elze, J., Bambach, M.: Simulation of different path strategies for wire-arc additive manufacturing with Lagrangian finite element methods. In: LS-DYNA Forum (2018)
38. Ito, M., Izawa, S., Fukunishi, Y., Shigeta, M.: Sph simulation of gas arc welding process. In: Proceeding Seventh International Conference on Computational Fluid Dynamics (Hawaii, 2012), ICCFD7-3706 (2011)
39. Jaffe, A.: The Millennium Grand Challenge in Mathematics. Notices of the AMS **53**(6), 652–660 (2006)
40. Jiang, J., Ma, Y.: Path planning strategies to optimize accuracy, quality, build time and material use in additive manufacturing: a review. Micromachines **11**(7), 633 (2020)
41. Kantorovich, L.: Mathematical methods of organizing and planning production. Manag. Sci. **6**(4), 366–422 (1960). English translation of the original Russian paper from 1939

42. Kim, F.H., Moylan, S.P.: Literature review of metal additive manufacturing defects. Tech. rep., U.S. Department of Commerce (2018)
43. Kolossov, S., Boillat, E., Glardon, R., Fischer, P., Locher, M.: 3D FE simulation for temperature evolution in the selective laser sintering process. Int. J. Mach. Tools Manuf. **44**(2), 117–123 (2004)
44. Koopmans, T.: Exchange Ratios between Cargoes on Various Routes (Non- Refrigerated Dry Cargoes). Tech. rep, Memorandum for the Combined Shipping Adjustment Board, Washington, D.C. (1942)
45. Kotadia, H., Gibbons, G., Das, A., Howes, P.: A review of laser powder bed fusion additive manufacturing of aluminium alloys: microstructure and properties. Addit. Manuf. **46**, 102155 (2021)
46. Kotzem, D., Dumke, P., Sepehri, P., Tenkamp, J., Walther, F.: Effect of miniaturization and surface roughness on the mechanical properties of the electron beam melted superalloy Inconel®718. Prog. Addit. Manuf. **5**(3), 267–276 (2019)
47. Körner, C., Bauereiß, A., Attar, E.: Fundamental consolidation mechanisms during selective beam melting of powders. Model. Simul. Mater. Sci. Eng. **21**(8), 085011 (2013)
48. Kucharik, M., Liska, R., Vachal, P., Shashkov, M.: Arbitrary Lagrangian-Eulerian (ALE) methods in compressible fluid dynamics. Programs and Algorithms of Numerical Mathematics **13**, 178–183 (2006)
49. Land, A., Doig, A.: An automatic method for solving discrete programming problems. Econometrica **28**, 497–520 (1960)
50. Leyffer, S.: Mixed-integer PDE-constrained optimization. In: Liberti, L., Sager, S., Wiegele, A. (eds.) Mixed-integer Nonlinear Optimization: A Hatchery for Modern Mathematics, vol. 46, pp. 2738–2740. Mathematisches Forschungsinstitut Oberwolfach, Germany (2015)
51. Liu, M., Chiu, L.N., Vundru, C., Liu, Y., Huang, A., Davies, C., Wu, X., Yan, W.: A characteristic time-based heat input model for simulating selective laser melting. Addit. Manuf. **44**, 102026 (2021)
52. Motzkin, T.: Beiträge zur Theorie der Linearen Ungleichungen. Ph.D. thesis, University of Zurich, Zurich (1936)
53. Nemhauser, G., Wolsey, L.: Integer and Combinatorial Optimization. Wiley, New York (1988). https://doi.org/10.1002/9781118627372
54. Nguyen, L., Buhl, J., Bambach, M.: Continuous Eulerian tool path strategies for wire-arc additive manufacturing of rib-web structures with machine-learning-based adaptive void filling. Addit. Manuf. **35**, 101265 (2020)
55. Nguyen, L., Buhl, J., Israr, R., Bambach, M.: Analysis and compensation of shrinkage and distortion in wire-arc additive manufacturing of thin-walled curved hollow sections. Addit. Manuf. **47**, 102365 (2021)
56. Padberg, M., Rinaldi, G.: A branch-and-cut algorithm for the resolution of large-scale symmetric traveling salesman problems. SIAM Rev. **33**(1), 60–100 (1991)
57. Papazoglou, E.L., Karkalos, N.E., Karmiris-Obratański, P., Markopoulos, A.P.: On the modeling and simulation of SLM and SLS for metal and polymer powders: a review. In: Archives of Computational Methods in Engineering (2021)
58. Persenot, T., Burr, A., Martin, G., Buffiere, J.Y., Dendievel, R., Maire, E.: Effect of build orientation on the fatigue properties of as-built electron beam melted Ti-6Al-4V alloy. Int. J. Fatigue **118**, 65–76 (2019)
59. Promoppatum, P., Uthaisangsuk, V.: Part scale estimation of residual stress development in laser powder bed fusion additive manufacturing of Inconel 718. Finite Elem. Anal. Des. **189**, 103528 (2021)
60. Ramos, D., Belblidia, F., Sienz, J.: New scanning strategy to reduce warpage in additive manufacturing. Addit. Manuf. **28**, 554–564 (2019)
61. Richardson, L.: The approximate arithmetical solution by finite differences of physical problems involving differential equations with an application to the stress in a masonry dam. Philos. Trans. R. Soc. Lond. A **210**, 307–357 (1911)

62. Rothberg, E.: An evolutionary algorithm for polishing mixed integer programming solutions. INFORMS J. Comput. **19**, 534–541 (2007)
63. Scherillo, F.: Chemical surface finishing of AlSi10mg components made by additive manufacturing. Manuf. Lett. **19**, 5–9 (2019)
64. Stern, F., Kleinhorst, J., Tenkamp, J., Walther, F.: Investigation of the anisotropic cyclic damage behavior of selective laser melted AISI 316L stainless steel. Fatigue Fract. Eng. Mater. Struct. **42**(11), 2422–2430 (2019)
65. Szost, B.A., Terzi, S., Martina, F., Boisselier, D., Prytuliak, A., Pirling, T., Hofmann, M., Jarvis, D.J.: A comparative study of additive manufacturing techniques: residual stress and microstructural analysis of CLAD and WAAM printed Ti–6Al–4V components. Mater. Des. **89**, 559–567 (2016)
66. Taler, J., Ocłoń, P.: Finite element method in steady-state and transient heat conduction. Encyclopedia of Thermal Stresses **4**, 1604–1633 (2014)
67. Virtanen, P., Gommers, R., Oliphant, T.E., Haberland, M., Reddy, T., Cournapeau, D., Burovski, E., Peterson, P., Weckesser, W., Bright, J., van der Walt, S.J., Brett, M., Wilson, J., Millman, K.J., Mayorov, N., Nelson, A.R.J., Jones, E., Kern, R., Larson, E., Carey, C.J., Polat, İ., Feng, Y., Moore, E.W., VanderPlas, J., Laxalde, D., Perktold, J., Cimrman, R., Henriksen, I., Quintero, E.A., Harris, C.R., Archibald, A.M., Ribeiro, A.H., Pedregosa, F., van Mulbregt, P., SciPy 1.0 Contributors: SciPy 1.0: fundamental algorithms for scientific computing in Python. Nat. Methods **17**, 261–272 (2020)
68. Wolsey, L.: Integer Programming, 2nd edn. Wiley, New York (2020)
69. Zhang, Y., Chen, Q., Guillemot, G., Gandin, C.A., Bellet, M.: Numerical modelling of fluid and solid thermomechanics in additive manufacturing by powder-bed fusion: continuum and level set formulation applied to track- and part-scale simulations. Comptes Rendus Mécanique **346**(11), 1055–1071 (2018)
70. Zhang, Y., Chen, X., Jayalakshmi, S., Singh, R.A., Deev, V.B., Prusov, E.S.: Factors determining solid solution phase formation and stability in CoCrFeNiX0.4 (X = Al, Nb, Ta) high entropy alloys fabricated by powder plasma arc additive manufacturing. J. Alloys Compd. **857**, 157625 (2021)

Unsupervised Optimization of Laser Beam Trajectories for Powder Bed Fusion Printing and Extension to Multiphase Nucleation Models

Ashkan Mansouri Yarahmadi, Michael Breuß, Carsten Hartmann, and Toni Schneidereit

Abstract In laser powder bed fusion, it is known that the quality of printing results crucially depends on the temperature distribution and its gradient over the manufacturing plate. We propose a computational model for the motion of the laser beam and the simulation of the time-dependent heat evolution over the plate. For the optimization of the laser beam trajectory, we propose a cost function that minimizes the average thermal gradient and allows to steer the laser beam. The optimization is performed in an unsupervised way. Specifically, we propose an optimization heuristic that is inspired by the well-known traveling salesman problem and that employs simulated annealing to determine a nearly optimal pathway. By comparison of the heat transfer simulations of the derived trajectories with trajectory patterns from standard printing protocols we show that the method gives superior results in terms of the given cost functional.

Keywords Additive manufacturing · Multiphase alloys · Trajectory optimization · Powder bed fusion printing · Heat simulation · Linear-quadratic control

1 Introduction

Nowadays, fabrication of individualized components made out of important engineering materials is possible with the help of Additive Manufacturing (AM), whose general strategy consists of building-up a preferably dense workpiece layer-by-

The authors acknowledge funding of their research by the European Regional Development Fund, EFRE 85037495.

A. Mansouri Yarahmadi (✉) · M. Breuß · C. Hartmann · T. Schneidereit
BTU Cottbus-Senftenberg, Institute for Mathematics, Cottbus, Germany
e-mail: yarahmadi@b-tu.de; breuss@b-tu.de; hartmanc@b-tu.de; schneidereit@b-tu.de

E. Cristiani et al. (eds.), *Mathematical Methods for Objects Reconstruction*, Springer INdAM Series 54, https://doi.org/10.1007/978-981-99-0776-2_6

layer [15]. Laser powder bed fusion (L-PBF) techniques use a deposited powder bed which is selectivity fused by a computer-controlled laser beam [36]. In general, a variety of different laser beam parameters such as laser power, scan speed, building direction and laser thickness influence the final properties of the fabricated product. A recent work [36] sheds light on the importance of the powder properties, namely the flow-ability, particle shape, and thermal properties of the powder mixture of spherical iron particles with mechanically crushed *ferroalloy* particles.

The term ferroalloy refers to various alloys of iron with a high proportion of one or more other elements such as manganese, aluminum, or silicon. It is used in production of steel and introduces distinct qualities, such as strength, ductility, and fatigue or corrosion resistance [7], to steel during production, and therefore it is closely associated with the iron and steel industry [35]. As an example, one can consider the production of ultra-low carbon steel [42], at which various ferroalloys can be added into the steel during different stages of ladle treatment [37], as the ultra-low carbon steel can be used in automobile industry [41].

One of the motivations behind using ferroalloy is the possibility to compensate in general a higher cost of AM by using the pre-alloyed powders available in market. This highly reduces the production cost compared to the case with a need to investigate a suitable mixture of metal alloys, which only few studies have reported [4, 30, 33]. In general, the development of new alloys for L-PBF is not a simple task due to the limiting final part quality and, usually, it requires a high economical investment that ultimately increases the price of final parts [12, 34]. However, as an emerging field of work [20, 28], additive manufacturing of soft magnetic core materials fabricated from ferroalloys using L-PBF techniques may provide new opportunities in the development of next-generation, high-performance energy converters. In clean energy technologies, such as wind power and pure electric power-trains, efficient electric motors use soft magnets fabricated from iron and silicon (Fe-Si) ferroalloy in the stator part.

Due to intense thermal input during additive manufacturing, the printed product can have defects, such as high deviations from the target geometry or cracks caused by large temperature gradients. For example, inhomogeneous heating may lead to unmelted powder particles that can locally induce pores and microscopic cracks [13]. In [21], a structural optimization problem is formulated which takes a control of porosity effect as a constraint in account aiming to obtain an optimal robust design with respect to the emergence of small holes during the manufacturing process. The key ingredient used in [21] is the concept of topological derivative, calculated and incorporated into a level set based shape optimization method developed based on the original idea proposed by Osher and Sethian [25]. In addition, the cooling process determines the microstructure of the printed workpiece and thus material properties, such as strength or toughness, which depend on the proportion of carbon embedded in the crystal structure of the material [2]. For crack-sensitive alloys and components with complex shapes, high strengths can be achieved through rapid cooling (bainitization), and one goal of this study is to devise a framework that may finally allow for controlling the phase transitions between the crystalline phases of the material through targeted scanning of the component during laser melting, thus avoiding the formation of microstructures (specifically: ferrite and pearlite) that are unfavorable for the strength of the material [17].

When modelling the cooling process, physical effects such as heat conduction and phase separation must be taken into account. To calculate laser controls for a realistic size of the components, grid-based numerical methods, in which the computational effort increases exponentially with the number of degrees of freedom can be prohibitive. As far as only the forward simulation of the AM process is concerned, grid-based schemes, such as finite element, finite difference, or finite volume methods, can be used, but in the context of filtering or control problems that may require to solve inverse problems or two-point boundary value problems, grid-based schemes are often prohibitively expensive. The problem is even more severe when the printed workpiece has a complicated shape that involves vastly different length scales (e.g., due to irregular and thin, elongated structures that are attached to large scale structures) that give rise to different temperature equilibration time scales and that then have to be resolved by local mesh-refinement in space and time or appropriate multilevel techniques [6]. The numerical complexity of classical numerical schemes currently prevents real-time algorithms to simulate and control the AM processes.

In this paper, we therefore aim at devising heuristic control strategies for the AM processes that allow for tight temperature control inside a workpiece, such that the cooling rate is controlled in order to maintain a target phase composition, while the temperature distribution is kept uniform during cooling, so as to avoid stress cracks. With these insights, we opt for investigating the laser beam trajectory as one of the key parameters in manufacturing costs of L-PBF. Specifically, we aim at conducting controlled laser beam simulation that approximately achieves *temperature constancy* on a melted power bed. In context of trajectory optimization, one can also refer to the works in [1, 5]. In [5], a combined temperature and phase-based objective function is devised and minimized aiming to result an optimum laser path. Another mixed cost function consists of spatio-temporal and thermal constraints is also addressed in [1] by considering a discrete version of an additive manufacturing problem. The discretization reduces the problem into a Mixed Integer Linear Programming (MILP) problem, which was solved using off-the-shelf solvers [14].

In our current study we mainly focus on the temperature constancy as an underlying assumption so that the nucleation process can be monitored through an appropriate phase kinetics model that is an extension of the famous JMAK model proposed by Johnson and Mehl [18], Avrami [3], and Kolmogorov [19]. The model allows to monitor the time-dependent phase transformation of the material under continuous cooling if one ignores the dependence of thermal conductivity of the phase composition. Let us also briefly highlight the nucleation as the process of transforming the austenite to ferrite during cooling, affecting the morphology of ferrite with a high impact on mechanical properties of the steel [44].

Outline The article is structured as follows: In Sect. 2, we first describe the heat transfer model and propose a cost function that consists of two terms aiming to maintain an almost constant temperature with a low spatial gradient across the power bed area. For simplicity, we confine ourselves to a 2-dimensional domain. In Sect. 2.2, we explain the idea of the travelling salesman problem (TSP) heuristics for the laser beam protocol; being one of the most fundamental and well-studied NP-hard problem in field of combinatorial optimization (e.g., [9, 11]), we will use a stochastic optimization strategy (simulated annealing) to compute protocols that are approximately optimal. In Sect. 3, we present extensive numerical tests for an Fe-Si ferroalloy, comparing temperature variation maps of TSP based protocols and three different standard protocols from L-PBF, namely *row-wise*, *zig-zag*, and *spiral* movements of the laser beam over the probe. Finally, the observations are summarized in Sect. 4 that also contains a discussion of possible extensions of the classical JMAK model to diffusion-controlled phase transformation with temperature control and a discussion of strategies to couple it to the heat transfer model; for the reader's convenience, we give a brief derivation of the original JMAK model in Appendix A.

2 Heat Transfer Model and TSP Formulation

As indicated, we first describe our heat simulation setting which is the framework for the TSP optimization protocol described in the second part of this section.

2.1 Heat Simulation Framework

We set up a simulation environment, namely *(i)* a moving source of heat (cf. (3)) to act as a laser beam on *(ii)* an area $\Omega = [-1, +1]^2 \subset \mathbb{R}^2$ simulated as deposition of (Fe-Si) metal powder called a plate which is assumed to be set by a constant Dirichlet boundary value; we assume that the plate is mounted to a baseplate with large thermal conductivity, which makes the choice of Dirichlet boundary conditions appropriate, since the base plate approximately acts as an (infinite) heat reservoir with constant temperature; if the surrounding is an insulator, then a reflecting, i.e., zero-flux or von Neuman boundary condition is more suitable. A sequence of laser beam movements called a trajectory that we dynamically obtain as a subset of sparse and equidistant 16×16 stopping points spanned over Ω, is followed so that at each sparse point the heat Eq. (1) is resolved based on a finite element method (FEM) [8] providing us with a temperature map that varies on different plate locations as the time evolves. At each stopping point one temperature map of size 100×100 pixels is obtained by interpolating [24] the produced solution on mesh level of the heat equation solved by FEM. The motivation to apply the laser beam only on a 16×16 subset points of Ω and not on all of it is to reduce the time complexity of our proposed simulation.

Letting u be the temperature at $(x, y) \in \Omega$ at time $t \in [0, T]$, the heat equation that governs the time evolution of u reads

$$\frac{\partial}{\partial t}u(x, y, t) = \alpha\nabla^2 u(x, y, t) + \beta I(x, y, t), \quad (x, y, t) \in \Omega^\circ \times (0, T) \tag{1a}$$

$$u(x, y, t) = \theta_0 \quad (x, y, t) \in \partial\Omega \times [0, T] \tag{1b}$$

$$u(x, y, 0) = u_0(x, y) \quad (x, y) \in \Omega, \tag{1c}$$

where we denote by

$$\nabla^2\phi = \frac{\partial^2\phi}{\partial x^2} + \frac{\partial^2\phi}{\partial y^2} \tag{2}$$

the Laplacian of some function $\phi \in C^2$, by Ω° we denote the interior of the domain Ω, and by $\partial\Omega$ its piecewise smooth boundary; here u_0 is some initial heat distribution, θ_0 is the constant ambient space temperature (20°C), and we use the shorthands

$$\alpha := \frac{\kappa}{c\rho} \quad \text{and} \quad \beta := \frac{1}{c\rho}$$

with κ, c, and ρ all in $\mathbb{R}^+$, denoting thermal conductivity, specific heat capacity, and mass density of the powder and the fused materials. For simplicity, we assume that the material parameters are constant. Even though, this assumption is not essential for the proposed setting, but it considerably simplifies the optimization problems. In practice, the thermal conductivity will not be constant but rather depends on shape, layer number, and more importantly temperature. Nevertheless the assumption that it is constant serves as a starting point to simplify the optimization and is justified by the fact that we do not consider solid-liquid phase transitions here during which the thermal conductivity may quickly drop [10]. The power density distribution of choice to simulate a laser beam is a Gaussian function:

$$I(x, y, t) = I_0 \cdot \exp\left\{\left[-2\left[\left(\frac{x - x_c(t)}{\omega}\right)^2 + \left(\frac{y - y_c(t)}{\omega}\right)^2\right]\right]\right\} \tag{3}$$

with an intensity constant

$$I_0 = \frac{2P}{\pi\omega^2} \tag{4}$$

by knowing ω and P to be the radius of the Gaussian beam waist and the laser power, respectively.

In our study we let $u(x, y, 0) = 0$ and $t \in \mathbb{R}^+$ and consider the realistic (Fe-Si) thermal properties, namely $\kappa = 0.180\left(\frac{\text{W}}{\text{mm}\cdot\text{K}}\right)$, $c = 0.939\left(\frac{\text{kJ}}{\text{kg}\cdot\text{K}}\right)$ and $\rho = 2600 \times 10^{-9}\left(\frac{\text{kg}}{\text{mm}^3}\right)$ (see [43]).

We solved (1) using [26] by setting $P = 4200\,(\text{W})$ and $\omega = 35$ pixels while letting (x_c, y_c) to take all possible trajectory points such that the domain Ω be always affected by five consecutive heat source moves. The consecutive points comprise five neighbors on the trajectory and their location on board solely depends on the type of the adopted trajectory. The velocity of the nozzle move among these points is of importance concerning the microstructure of the printed product that we will discuss later. In this way and by letting each of these points to represent the center of the Gaussian heat source, we simulated the heat source movement across the board. For further clarification, let us stress that in this study we highly rely on [26] that operates on a mesh representation of the defined plate geometry on $\Omega = [-1, +1]^2 \subset \mathbb{R}^2$. To provide [26] a heat source, first we created a discrete format of Gaussian heat source with a radius of 35 pixel that was later triangulated over a mesh and fed to [26] to heat up the plate geometry in mesh level. It consequently means, the heat source is operating over the plate on mesh level, though we represent the obtained results on pixel level and image format.

A schematic view of the row-wise, the zig-zag, and the spiral protocols are visualized in Fig. 1. As indicated, the TSP [9] and its customization into our problem is clearly revealed with more details in Sect. 2.2 below.

Control Objective By adoption of these protocols shown as Fig. 1, we aim to see which one leads to a minimized value of a desired objective function:

$$J(m) = \frac{1}{2}\int_0^T \left(\int_\Omega |\nabla u_m(z,t)|^2 + \left(u_m(z,t) - u_g\right)^2 dz\right) dt\,, \tag{5}$$

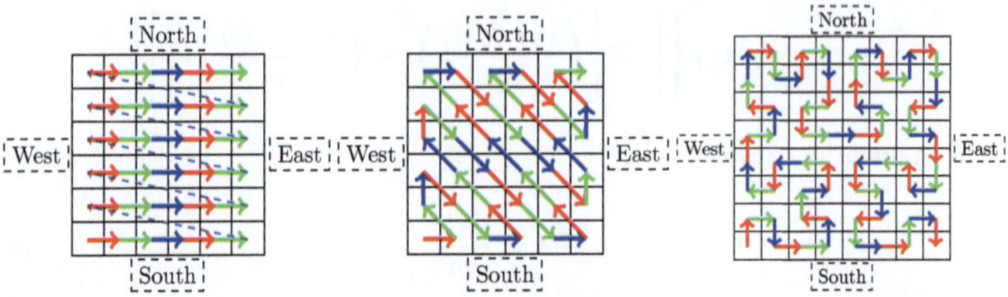

Fig. 1 Three different protocols used to arrange the movement of the heat source, **(left)** row-wise **(middle)** zig-zag and **(right)** spiral protocol (constructed by a Hilbert polygon). A movement always starts with a red arrow and proceeds in sequence of green, blue followed with another red arrow and this sequence of colors continues to repeat and construct the entire trajectory of each protocol

with

$$u_m = u_m(z,t)\,, \quad z = (x,y) \in \Omega\,, \,, t \in [0,T] \tag{6}$$

being the solution of the heat Eq. (1) on the interval $[0, T]$ under the control

$$m\colon [0,T] \to \Omega\,, \quad t \mapsto (x_c(t), y_c(t)) \tag{7}$$

that describes the trajectory of the center of the laser beam. Moreover, we have introduced u_g as the desired target temperature to be maintained over the domain Ω as time t evolves.

The motivation behind (5) is to maintain a smooth temperature gradient over the entire plate for all $t \in [0, T]$ as is achieved by minimizing the L^2-norm of the gradient, ∇u, while at the same time keeping the (average) temperature near a desired temperature u_g for any time t.

We proceed by dividing the entire plate Ω into 16 sub-domains (see Fig. 2) and investigate our objective function (5) within each sub-domain as explained in Sect. 3. The motivation behind having sub-domains is to maintain a temperature constancy within them by finding an optimum trajectory. In this way, we opt in our future work to observe the effect of nozzle trajectories on cooling process and phase transformation governed by JMAK model (18) within each sub-domain.

2.2 TSP Based Formulation

A common assumption among numerous variants of the TSP, as an NP-hard problem [11], is that a set of cities have to be visited on a shortest possible *tour*. Let $\mathcal{C}_{n\times n}$ be a symmetric matrix specifying the distances corresponding to the paths connecting a vertex set $\mathcal{V} = \{1, 2, 3, \cdots, n\}$ of cities to each other where $n \in \mathbb{N}$ is the number of cities. A tour over the complete undirected graph $\mathcal{G}(\mathcal{V}, \mathcal{C})$ is defined as a cycle passing through each vertex exactly once. The traveling salesman problem seeks a tour of minimum distance.

To adapt the TSP into our context, we formulate its input $\mathcal{V}$ as the set of all 16×16 stopping points of the heat source over the board Ω, and the set $\mathcal{C}$ as a penalty matrix that elements of $\mathcal{C}_{ij} \geq 0$ are the impact (i.e., cost) of moving the heat source from a $i \in \mathcal{V}$ to $j \in \mathcal{V}$. For every vertex $i \in \mathcal{V}$, the possible movements to all $j \neq i$ with the associated cost (5) is computed and assigned to $\mathcal{C}_{ij}$ (see below for details). With this formulation, we want to remind the reader that $\mathcal{C}$ elements are non-negative and follow the triangle inequality:

$$\mathcal{C}_{ij} \leq \mathcal{C}_{ik} + \mathcal{C}_{kj} \tag{8}$$

with $i, j, k \in \mathcal{V}$.

Note that, the matrix C is obtained based on a prior set of temperature maps produced using FEM without enforcing any particular protocol on them. It means, each of the 16×16 stopping points is heated up separately and one by one while the resultant heat maps are kept to be used in later stages and while adopting the adopted protocols. Our motivation is to investigate if the movement of the heat source based on a TSP may result in any improvement while minimizing (5) in contrast to the other three protocols.

With this general formulation at hand, let us have a closer look on the discretized form of (5) that was used in the current study to compute the elements of the penalty matrix:

$$c_{ij} = \left| \sum_{l=1}^{4\times4} \left(\|\boldsymbol{\Psi}(i,l)\|^2 + \left(\boldsymbol{\Lambda}(i,l) - u_g\right)^2 \right) - \sum_{l=1}^{4\times4} \left(\|\boldsymbol{\Psi}(j,l)\|^2 + \left(\boldsymbol{\Lambda}(j,l) - u_g\right)^2 \right) \right| \tag{9}$$

with l to be the sub-domain index. In addition, $\boldsymbol{\Psi}(\cdot,l) = \sum_{z\in\Omega_l}\sum_{t\in t_l} \nabla u_m(z,t)$ represents the temperature gradient aggregation within each sub-domain, and $\boldsymbol{\Lambda}(\cdot,l) = \frac{1}{|\Omega_l|}\sum_{z\in\Omega_l}\sum_{t\in t_l} u_m(z,t)$ is the average temperature value of each sub-domain, with t_l to be the time period for which the nozzle operates on Ω_l. Let us here shortly note few points on discretization of (9) based on (5). Though [26] is performed on mesh level over the entire Ω, it also provides us heat maps on pixel level that we use to compute (9) and within the divided sub-domains, namely all Ω_l. Here and by $|\Omega_l|$, we mean the number of pixel points in Ω_l. Since we rely on [26] to avoid our own mesh processing, the time period for which the nozzle operates on mesh points comprising Ω_l is not known for us. Thus, the term t_l is borrowed from mesh processing phase of our work supported by Partial Differential Equation Toolbox [26]. With this insights, (9) represents the TSP cost of moving the nozzle from the ith to the jth stopping point that depends on (a) the mean square deviation of the temperature field from constancy and (b) on the mean square deviation from the global target temperature u_g. In our simulation, the nozzle moves in the direction of the shortest (Euclidean) path connecting two successive stopping points where we assumed the nozzle always adjusts its velocity so that the path between any arbitrarily chosen stopping points i and j always takes the same amount of time. The motivation behind is to heat up successive stopping points i and j on same time interval and irrespective of their Euclidean distance. In a real scenario, one should be aware of the importance of a nozzle with flexible velocity in direction of maintaining a temperature constancy constraint. As strong heat flux from the recently heated up points into the already solidified parts leads to high heating and cooling rates, which promote the formation of cold cracks due to the high thermal stresses [4].

In practice no polynomial-time algorithm is known for solving the TSP [11], so we adopt a *simulated annealing algorithm* [39] that was first proposed in statistical physics as a means of determining the properties of metallic alloys at a given temperatures [22]. In TSP context, the simulated annealing starts by taking a tour as a random permutation of all the stopping points. Next, the taken tour is randomly perturbed and its obtained cost based on (9) is compared to a borrowed parameter from [39] called temperature that is set to an initial high value. As the number of iterations increases the so-called cooling schedule [39] decays the temperature parameter leading us to the final optimal tour. At each iteration, the newly perturbed tour is accepted if it has a lower cost compared to the temperature parameter. If the new tour has a higher cost, then it is accepted with a probability determined by the temperature parameter and the difference between the existing and the new costs.

In Sect. 3, we reveal our results obtained by adopting a TSP based protocol along with three other protocols. Now, let us observe in Sect. 3 a set of temperature maps obtained based on all four protocols.

3 Results

To obtain our data set, we let the simulated source of heat to move over Ω through a set of 16×16 stopping points located equidistant from each other in both x and y directions. In this way, we obtain a set of 256 temperature maps, each of size 100×100 number of pixels. Each temperature map is calculated globally but the cost function (5) computation is performed locally and within each of 4×4 predefined sub-domains. Figure 2 shows a set of consecutive moves of the heat source propagating heat globally across Ω, with the 16 sub-domains divided by white lines from each other.

Let us have a look at Figs. 3, 4, 5, and 6 corresponding to all four protocols, from each only three randomly chosen temperature maps are visualized. The sample maps are taken to show the variation of the diffused heat patterns across Ω. All temperature values in the plot are in °C.

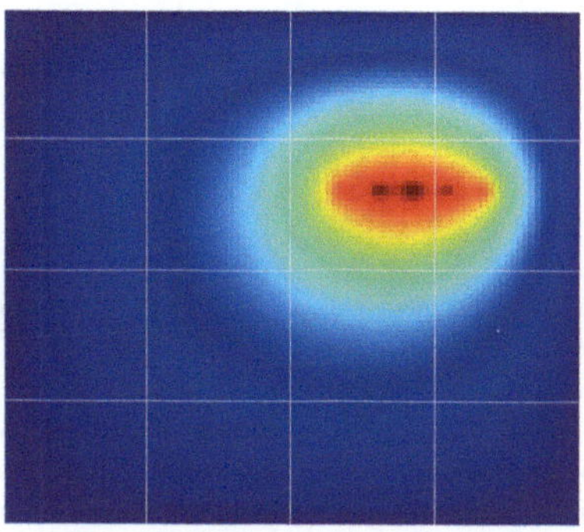

Fig. 2 The diffused temperature across the domain Ω heated up by a set of moving heat sources. We divide the entire domain into 4×4 sub-domains separated by white lines. Within each sub-domain, (5) is computed to reveal how the temperature gradient $|\nabla u(\cdot, t)|$ is diffused as a function of time t and how the average temperature $\bar{u}$ is maintained near to a target value of u_g. We applied four different protocols shown in Fig. 1 and computed our desired quantities within each sub-domain locally

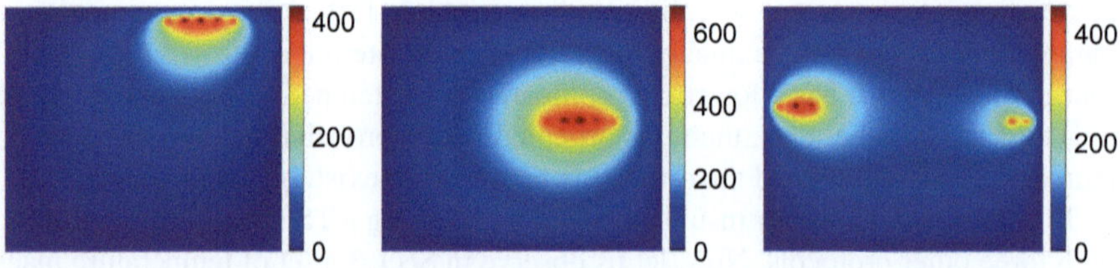

Fig. 3 Typical temperature maps obtained by enforcing the row-wise protocol

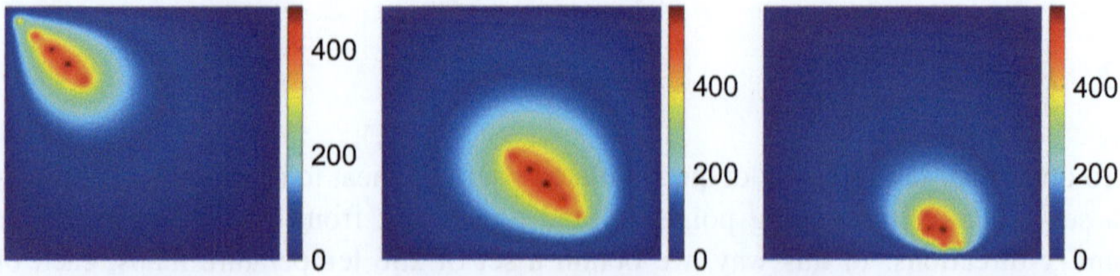

Fig. 4 Three typical temperature maps produced by the zig-zag protocol

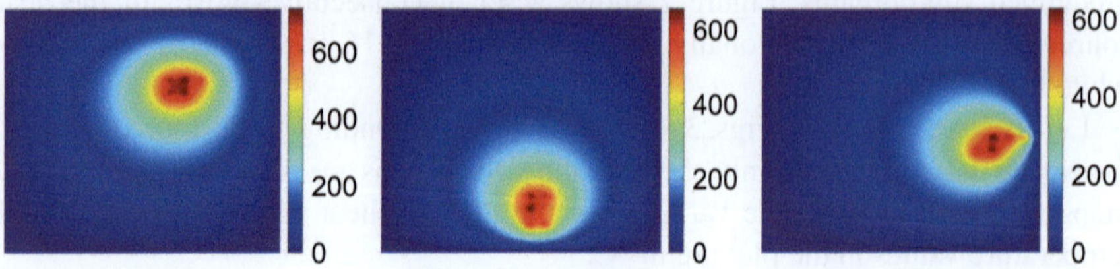

Fig. 5 Typical temperature maps generated by the spiral protocol

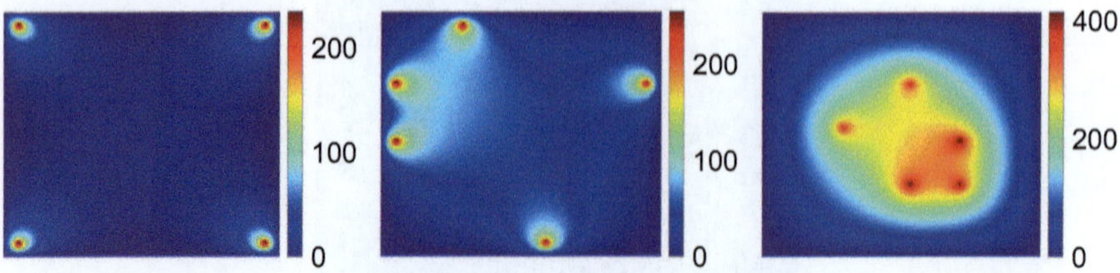

Fig. 6 Temperature maps based on the TSP solution. The most two left images clearly show that the second term of (5) contributes to the heat source locations so that an almost uniform temperature field is maintained across Ω. All temperature values in the plot are in °C

By only looking at Figs. 3, 4, 5, and 6, one may not be able to elaborate on variation details of (5) concerning each protocol within each sub-domain, but one may get a quick impression that the TSP based protocol provides more control so that a constant temperature can be maintained across Ω. The left two images of Fig. 6 specifically show the chosen locations of the heat source by the TSP based

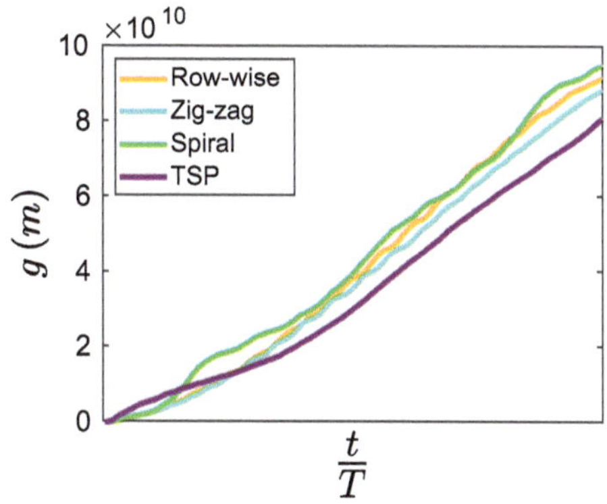

Fig. 7 The computed average cost function (5) concerning all protocols as time $t/T \in [0, 1]$ evolves. As one clearly observes, the heat source movement over the entire Ω through the 16×16 grid leads to the minimized version of the cost function in case the TSP protocol is followed. Note that, we run the whole process four times and computed the average cost values as visualized as current figure

protocol, intended to maintain almost a uniform temperature pattern across Ω. For simplicity, we have chosen the target temperature $u_g = 293.15\text{K} = 20°\text{C}$ to be maintained within each of the sub-domains that represents the environment temperature. This temperature tracking property is necessary in order to be able to control the phase composition. This is clearly in contrast to the results of the three other protocols that use no information whatsoever about the actual cost but perform only based on a predefined motion pattern.

Finally, we run the whole process four times and computed the average cost values concerning each protocol. As Fig. 7 visualizes and as we have already expected, the TSP protocol has the lowest average cost among all protocols.

4 Conclusion and Discussion

We have developed a heuristic for trajectory optimization for the use with powder bed fusion printing based on an approximate solution to a travelling salesman (TSP) problem. The aim was to determine printing trajectories that have favorable properties in terms of the heat distribution of the melt. As we confirm on the basis of a simplified heat transfer model, the proposed TSP optimization scheme robustly shows a slightly more homogeneous heat distribution, in comparison to what is obtained from standard laser beam patterns used in additive manufacturing. Let us emphasize that our approach is based on an open-loop control that does not take into account the actual heat distribution inside the printed material. To be more clear, we assert that the temperature distributions diffuse across the sub-domains and are not limited within each sub-domain. What sounds like a soft spot can actually be an advantage since (a) the full, spatially resolved temperature information may not be available during the printing process and (b) computing optimal feedback laser controls in real-time is likely to be impossible for 3D printing.

What is moreover missing is to include the transformation of the alloy phases during the cooling process. In Appendix B, we briefly discuss a possible extension of the framework to include simplified phase kinetics that can be used to optimize laser trajectories with regard to a homogeneous temperature distribution *and* a prescribed target phase composition of the printed material. We leave this for future work. Nevertheless the computational results for the simplified model already suggest that the trajectory optimization can indeed have a significant effect on the temperature distribution during the printing process, which indicates the potential usefulness of heuristic optimization tools in the field of 3D printing and suggests their implementation for powder bed fusion printing. Our approach takes into account that many manufacturers of printing devices prescribe trajectory patterns that can be computed offline; this is easily possible with the suggested TSP based optimization method.

A The Classical Johnson–Mehl–Avrami–Kolmogorov (JMAK) Model

We explain the idea behind the JMAK model that describes the volume fraction transformation of, e.g., *austenite* into *ferrite*. To this end, let $\Omega \subset \mathbb{R}^3$ be a bounded domain with reference volume $\mathcal{V}_r = |\Omega|$ that is called the *parent phase* (e.g., austenite), here called A phase. Consider another phase (e.g., ferrite) that we call the F phase that takes up a volume $\mathcal{V}_F$ at time $t > 0$ and that is contained inside Ω. The volume of the original, yet untransformed phase is denoted by $\mathcal{V}_A$ where $\mathcal{V}_A + V_F = \mathcal{V}_r$.

We seek an equation that describes the volume fraction

$$\mathcal{P}(t) = \frac{\mathcal{V}_F(t)}{\mathcal{V}_r}, \quad t > 0 \tag{10}$$

as a function of time. To this end, suppose that the volume of the F phase between time t and $t + dt$ changes by either growth of the existing phase or by formation (and growth) of new F nuclei (also: grains).

Clearly, only those parts of the F phase that lie in the previously untransformed phase A can contribute to $\mathcal{V}_F$ at time $t+dt$, so let $\mathcal{V}_F^{\mathrm{e}}$ at time $t+dt$ be the volume that $\mathcal{V}_F$ could have evolved into if there were no other F parts that could impinge upon one another. Assuming spatial homogeneity of the growing nuclei, the probability of impingement is given by $\mathcal{P} = \mathcal{V}_F/\mathcal{V}_r$, and thus, on average, the infinitesimal change in the volume is given by the famous Avrami equation [3]

$$d\mathcal{V}_F(t) = \mathcal{V}_r d\mathcal{P}(t) = (1 - \mathcal{P}(t))\, d\mathcal{V}_F^{\mathrm{e}}(t)\,. \tag{11}$$

We can integrate (11) to obtain $\mathcal{P}$ as a function of time: since (11) is equivalent to

$$\frac{1}{1-\mathcal{P}(t)}d\mathcal{P}(t) = \frac{d\mathcal{V}_F^{\mathrm{e}}(t)}{\mathcal{V}_r}, \tag{12}$$

it readily follows that

$$\mathcal{P}(t) = 1 - C\exp\left(-\frac{\mathcal{V}_F^{\mathrm{e}}(t)}{\mathcal{V}_r}\right), \tag{13}$$

where $C \in (0, 1]$ depends on the initial condition via $C = 1-\mathcal{P}(0)$. The problem is that we do not know how the extended volume evolves as a function of time, i.e., the equation for the volume fraction $\mathcal{P}(t)$ is not closed.

In order to close the equation, we have to model the growth of the F phase and the nucleation of new particles. The following assumptions are key to the famous (JMAK) model by Johnson and Mehl [18], Avrami [3], and Kolmogorov [19]:

1. Nucleation occurs randomly and homogeneously over the parent phase.
2. The growth rate does not depend on the fraction of the transformed phase.
3. Growth occurs at the same rate in all directions.

As a consequence, the transformation of A into F is governed by the nucleation of new particles and by their growth into spherical particles until they impinge upon each other.

We assume that the phase transformation occurs at constant temperature $\theta > 0$, and we call $\mu = \mu(\theta)$ the nucleation rate of F particles (i.e., the number of particles per unit time and volume) and $\rho(t) = \rho(\theta, t)$ is the diameter growth rate of the resulting spherical particles (i.e., length per unit time). For an isolated particle born at time $\tau \in (0, t]$, the volume occupied inside the parent phase at time t is given by (assuming that no impingement happens between τ and t)

$$v(t,\tau) = \frac{4\pi}{3}\left(\int_\tau^t \rho(s)\,ds\right)^3 . \tag{14}$$

Assumption (1) implies that particles are randomly generated by a Poisson point process, i.e., the probability that exactly k particles or grains are randomly generated within a bounded region, say, $B \subset \mathcal{V}$ is equal to (e.g., see [38, Ch. V])

$$\mathbb{P}(k \text{ new particles inside } B) = \frac{(\mu|B|)^k}{k!}e^{-\mu|B|}. \tag{15}$$

Since particles are generated at a constant rate μ, the average extended (i.e., hypothetical) volume occupied at time t is given by

$$\mathcal{V}_F^{\mathrm{e}}(t) = \mu\mathcal{V}_r \int_0^t v(t,\tau)\,d\tau . \tag{16}$$

Inserting the last two equations into (13) and assuming $\mathcal{P}(0) = 0$, we obtain

$$\mathcal{P}(t) = 1 - \exp\left(-\frac{4\pi\mu}{3}\int_0^t \left(\int_\tau^t \rho(s)\, ds\right)^3 d\tau\right). \tag{17}$$

Note that the last equation is not closed as it depends on the unknown diameter growth rate ρ. The classical JMAK equation is now obtained from (17) when $\rho(s) = \rho_0$ is constant, which leads to

$$\mathcal{P}(t) = 1 - \exp\left(-\mathcal{K}t^4\right) \tag{18}$$

with the material constant $\mathcal{K} = \dfrac{\pi\mu\rho_0^3}{3}$ that combines the contributions of nucleation rate μ and growth rate ρ_0.

B Extension to Multiphase Alloys and Rapid Cooling

We will now discuss the effect of phase transformation upon cooling. More specifically, we assume that the material under consideration undergoes a transformation from its (e.g., austenitic) parent phase to another (e.g., ferritic) phase. Such a phase transformation is temperature-activated process that lead to significant changes of the microstructure of the materials during the cooling process.

Diffusion-Controlled Phase Transformation

The classical JMAK model, Eq. (18), is the result of simply assuming that the growth rate $\rho(s) = \rho_0$ is constant (see Appendix A). A drawback of the JMAK model (18) is that it is not easily applicable to situations in which the temperature is not constant during a phase transformation.

An alternative closure of (17) that has been suggested in [16] can be obtained by assuming that the growth rate of metallic phases depends on the carbon concentration inside the metal. The growth of the F phase then stops when a certain equilibrium of carbon concentration $\bar{C}_A = \bar{C}_A(\theta)$ is reached inside the parent phase. Let $C_A(t)$ the carbon concentration inside the parent (austenite) phase at time $t > 0$. We further assume that the carbon concentration inside the growing (ferrite) phase is constant and equal to its equilibrium value $\bar{C}_F = \bar{C}_F(\theta)$. By conservation of mass, the (total) background carbon concentration, C, satisfies

$$C = \bar{C}_F\mathcal{P}(t) + C_A(t)(1 - \mathcal{P}(t)). \tag{19}$$

In equilibrium, $\mathcal{P}(t) = \bar{\mathcal{P}}$ and $C_A(t) = \bar{C}_A$, and mass conservation is equivalent to

$$\bar{\mathcal{P}} = \frac{C - \bar{C}_A}{\bar{C}_F - \bar{C}_A}. \tag{20}$$

Under the assumption that the growth rate $\rho(t)$ is proportional to the difference of the time-dependent carbon concentration in the A phase (cf. [40, Eqn. (3)]), i.e.,

$$\rho(t) \propto \eta(t) \frac{\bar{\mathcal{P}} - \mathcal{P}(t)}{1 - \mathcal{P}(t)}, \tag{21}$$

with $\eta(t) = Mt^{-\gamma}$ for some suitable $\gamma \in (0, 1)$ and the coefficient $M = M(\theta) > 0$ describing the mobility (i.e., the square root of the diffusion coefficient) of carbon inside the A phase. We can eliminate the background carbon concentration C by combining (19) and (20), which then together with the last equation implies

$$\rho(t) = \frac{M}{\sqrt{t}} \frac{\bar{\mathcal{P}} - \mathcal{P}(t)}{1 - \mathcal{P}(t)}. \tag{22}$$

Note that the equilibrium phase fraction $\bar{\mathcal{P}} = \bar{\mathcal{P}}(\theta, C)$ depends on both temperature and the background carbon concentration inside the probe. Differentiating (17) and inserting (22) we obtain

$$\mathcal{P}'(t) = 4\pi \frac{\mu M}{\sqrt{t}} (\bar{\mathcal{P}} - \mathcal{P}(t)) \int_0^t \left(\int_\tau^t \frac{M}{\sqrt{s}} \frac{\bar{\mathcal{P}} - \mathcal{P}(s)}{1 - \mathcal{P}(s)}\, ds \right)^2 d\tau\,, \quad \mathcal{P}(0) = 0\,, \tag{23}$$

which is an integro-differential equation (IDE) for the phase fraction.

Mean-Field Coupling of Phase Kinetics and Heat Conduction The thermal conductivity of alloys depends on both phase composition and microstructure, where for simplicity we may assume that the thermal conductivity of the melted metal powder is an increasing affine function of $\mathcal{P}$ [27]. Since our phase kinetics model has no spatial resolution, the coupling to the heat equation requires to couple an ODE and a PDE model. A naive approach consists of letting the two models interact through their mean field. To this end, we define the average temperature

$$\bar{u}(t) = \int_\Omega u(x, y, t)\, w(x, y)\, dxdy\,, \tag{24}$$

where the weight function $w \geq 0$ is a probability density supported on Ω, e.g., the uniform density $w = |\Omega|^{-1}\mathbf{1}_\Omega$. Letting

$$\mathcal{P}(t) = \mathcal{P}(t; \bar{u}(t)) \tag{25}$$

denote the solution of (23) for $\mu = \mu(\bar{u}(t))$ and $M = M(\bar{u}(t))$, we end up with the following nonlinear modification of the heat equation (1):

$$\frac{\partial u}{\partial t} = \alpha(\mathcal{P})\nabla^2 u + L\frac{\partial \mathcal{P}}{\partial t} + \beta I\,, \qquad (x, y, t) \in \Omega^\circ \times (0, T) \tag{26a}$$

$$u = \theta_0 \qquad (x, y, t) \in \partial\Omega \times [0, T] \tag{26b}$$

$$u = u_0 \qquad (x, y, t) \in \Omega^\circ \times \{0\} \tag{26c}$$

where, as before, we denote by Ω° the interior of the domain Ω and by $\partial\Omega$ its piecewise smooth boundary. Equation (26) is coupled to the IDE

$$\mathcal{P}' = 4\pi \frac{\mu(\bar{u})M(\bar{u})}{\sqrt{t}}(\bar{\mathcal{P}} - \mathcal{P}) \int_0^t \left(\int_\tau^t \frac{M(\bar{u})}{\sqrt{s}} \frac{\bar{\mathcal{P}} - \mathcal{P}(s)}{1 - \mathcal{P}(s)}\, ds \right)^2 d\tau\,, \; \mathcal{P}(0) = 0 \tag{27}$$

for the phase kinetics. In comparison with the linear heat equation (1), the nonlinear equation (26) contains two extra terms:

(1) The phase dependent coefficient $\alpha \propto \kappa$ that depends on $\mathcal{P}$ by virtue of the thermal conductivity $\kappa(\mathcal{P}) = \kappa_0 - \kappa_1\mathcal{P}$.
(2) The term $L\partial\mathcal{P}/\partial t$ that describes the release of heat due to the phase transformation, with $L > 0$ having the physical unit of temperature.

We expect the system of Eqs. (26)–(27) is well-posed since both α and $\partial\mathcal{P}/\partial t$ are smooth and bounded. (We leave the analysis of the problem for future research.) To model rapid cooling of the probe to obtain a desired phase composition at some terminal time T, the cost functional (5) can be augmented by a (e.g., quadratic) terminal cost that penalizes deviations from the target phase composition at time T. For feedback control problems, in which the laser beam is driven according to the actual (measured) temperature field, an additional penalization term for the laser beam may be added to avoid unphysical moves.

We leave the detailed analysis of the model and its numerical simulation that also requires estimation of the temperature-dependent model parameters, such as $\kappa(\cdot)$ or $\mu(\cdot)$ from experimental data, for future work.

Illustration of the Phase Field Kinetics Equation (23) can be turned into an ODE system by substituting the variables

$$w(t) = \mu t\,,\; x(t) = \mu \int_0^t \left(\int_\tau^t \rho(s)\, ds \right) d\tau\,,\; y(t) = \mu \int_0^t \left(\int_\tau^t \rho(s)\, ds \right)^2 d\tau\,, \tag{28}$$

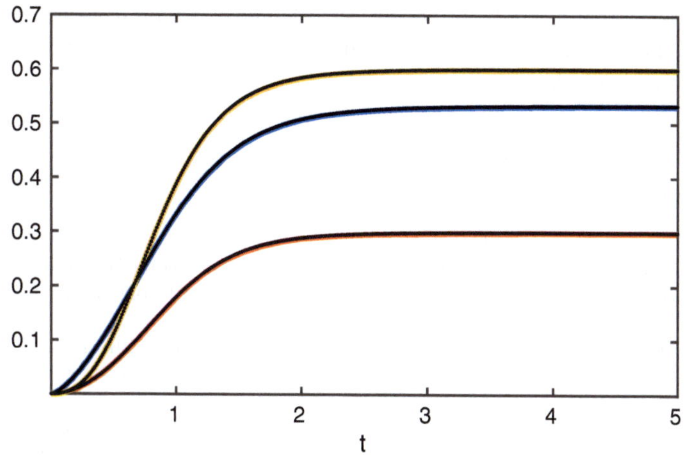

Fig. 8 Numerical solution to the system of Eqs. (29) with the parameters $\mu = 1$, $M = 1$, $\bar{\mathcal{P}} = 0.6$ (arbitrary units): x (orange), y (blue), and $\mathcal{P}$ (yellow). The dots show the SCNF solution, for comparison the colored lines show the numerical solution computed by a 4-th order Runge-Kutta scheme. The solution converges to the target phase fraction corresponding to the chosen temperature

with ρ as given by (22). It follows that the IDE (23) is equivalent to the following ODE system

$$w'(t) = \mu\,, \qquad w(0) = 0 \tag{29a}$$

$$x'(t) = \frac{M}{\sqrt{t}} \frac{\bar{\mathcal{P}} - \mathcal{P}(t)}{1 - \mathcal{P}(t)} w(t)\,, \qquad x(0) = 0 \tag{29b}$$

$$y'(t) = \frac{2M}{\sqrt{t}} \frac{\bar{\mathcal{P}} - \mathcal{P}(t)}{1 - \mathcal{P}(t)} x(t)\,, \qquad y(0) = 0 \tag{29c}$$

$$\mathcal{P}'(t) = \frac{4\pi M}{\sqrt{t}} (\bar{\mathcal{P}} - \mathcal{P}(t)) y(t)\,, \qquad \mathcal{P}(0) = 0\,, \tag{29d}$$

where for convenience we have suppressed the temperature dependence of the coefficients μ, M, and $\bar{\mathcal{P}}$. Note that the solution is smooth at the origin. (The initial values for x, y and $\mathcal{P}$ should be interpreted as right-sided limits.) For illustration, Fig. 8 shows a typical solution of the phase kinetics.

The solution shown in Fig. 8 has been computed numerically using the Subdomain Collocation Neural Forms (SCNF) method from [31, 32] which represents a specific physics-informed neural network (PINN) approximation [23, 29] that solves a nonlinear regression problem for a neural network with the squared difference between left and right hand side as a loss functions. The idea of using the SCNF approximation to approximate the phase field kinetics rather than, say, any standard

ODE solver is that it can be relatively easily combined with deep neural network approximations of the nonlinear heat equation and, possibly, with the computation of an optimal control.

References

1. Afzal, Z., Prabhakar, P., Prabhakar, P. Optimal tool path planning for 3D printing with spatio-temporal and thermal constraints. In: 2019 Sixth Indian Control Conference (ICC), pp. 176–181 (2019)
2. Ali, M., Porter, D., Kömi, J., Eissa, M., El Faramawy, H., Mattar, T.: Effect of cooling rate and composition on microstructure and mechanical properties of ultrahigh-strength steels. J. Iron Steel Res. Int. **26**, 1350–1365 (2019)
3. Avrami, M., III.: Granulation, phase change, and micro-structure. J. Chem. Phys. **9**, 177 (1941)
4. Baqerzadeh Chehreh, A., Strauch, A., Großwendt, F., Röttger, A., Fechte-Heinen, R., Theisen, W., Walther, F.: Influence of Different Alloying Strategies on the Mechanical Behavior of Tool Steel Produced by Laser-Powder Bed Fusion. Materials **14**, 3344 (2021)
5. Boissier, M., Allaire, G., Tournier, C.: Additive manufacturing scanning paths optimization using shape optimization tools. Struct. Multidiscip. Optim. **61**, 2437–2466 (2020)
6. Borzì, A., Schulz, V.: Multigrid methods for PDE optimization. SIAM Rev. **51**(2), 361–395 (2009)
7. Bublik, S., Olsen, J., Loomba, V., Reynolds, Q., Einarsrud, K.: A review of ferroalloy tapping models. Metall. Mater. Trans. B **52**, 2038–2047 (2021)
8. Fish, J., Belytschko, T.: A First Course in Finite Elements, vol. 1. Wiley, New York (2007)
9. Flood, M.: The traveling-salesman problem. Oper. Res. **4**, 61–75 (1956)
10. Foteinopoulos, P., Papacharalampopoulos, A., Stavropoulos, P.: On thermal modeling of Additive Manufacturing processes. CIRP J. Manuf. Sci. Technol. **20**, 66–83 (2018)
11. Garey, M., Johnson, D.: Computers and intractability. In: A Guide to the Theory of NP-Completeness. W. H. Freeman & Co., New York (1990)
12. Gibson, I., Rosen, D., Stucker, B., Khorasani, M.: Materials for additive manufacturing. In: Additive Manufacturing Technologies, pp. 379–428 (2021)
13. Großwendt, F., Röttger, A., Strauch, A., Chehreh, A., Uhlenwinkel, V., Fechte-Heinen, R., Walther, F., Weber, S., Theisen, W.: Additive manufacturing of a carbon-martensitic hot-work tool steel using a powder mixture—Microstructure, post-processing, mechanical properties. Mater. Sci. Eng. A **827**, 142038 (2021)
14. Gurobi Optimization, LLC Gurobi Optimizer Reference Manual (2022). https://www.gurobi.com
15. Herzog, D., Seyda, V., Wycisk, E., Emmelmann, C.: Additive manufacturing of metals. Acta Mater. **117**, 371–392 (2016)
16. Hömberg, D., Patacchini, D., Sakamoto, K., Zimmer, J.: A revisited Johnson-Mehl-Avrami-Kolmogorov model and the evolution of grain-size distributions in steel. IMA J. Appl. Math. **82**(4), 763–780 (2017)
17. Hu, X., Nycz, A., Lee, Y., Shassere, B., Simunovic, S., Noakes, M., Ren, Y., Sun, X.: Towards an integrated experimental and computational framework for large-scale metal additive manufacturing. Mater. Sci. Eng. A **761**, 138057 (2019)
18. Johnson, W.A., Mehl, K.E.: Reaction Kinetics in Processes of Nucleation and Growth. Trans. Am. Inst. Min. Metall. Pet. Eng. **195**, 416 (1939)
19. Kolmogorov, A.: On the statistical theory of the crystallization of metals. In: Bulletin of the Academy of Sciences of the USSR, Mathematics Series, vol. 1, pp. 355–359 (1937)

20. Lamichhane, T., Sethuraman, L., Dalagan, A., Wang, H., Keller, J., Paranthaman, M.: Additive manufacturing of soft magnets for electrical machines–a review. Mater. Today Phys. **15**, 100255 (2020)
21. Martínez-Frutos, J., Allaire, G., Dapogny, C., Periago, F.: Structural optimization under internal porosity constraints using topological derivatives. Comput. Methods Appl. Mech. Eng. **345**, 1–25 (2019)
22. Metropolis, N., Rosenbluth, A., Rosenbluth, M., Teller, A., Teller, E.: Equation of state calculations by fast computing machines. J. Chem. Phys. **21**, 1087 (1953)
23. Mowlavi, S., Nabi, S.: Optimal control of PDEs using physics-informed neural networks. arXiv:2111.09880 (2021)
24. Oetken, G., Parks, T., Schussler, H.: New results in the design of digital interpolators. IEEE Trans. Acoust. Speech Signal Process. **23**(3), 301–309 (1975)
25. Osher, S., Sethian, J.: Fronts propagating with curvature-dependent speed: algorithms based on Hamilton-Jacobi formulations. J. Comput. Phys. **79**, 12–49 (1988)
26. Partial Differential Equation Toolbox. The MathWorks Inc., Natick (2020)
27. Peet, M.J., Hasan, H.S., Bhadeshia, H.K.D.H.: Prediction of thermal conductivity of steel. Int. J. Heat Mass Transf. **54**(11), 2602–2608 (2011)
28. Plotkowski, A., Pries, J., List, F., Nandwana, P., Stump, B., Carver, K., Dehoff, R.: Influence of scan pattern and geometry on the microstructure and soft-magnetic performance of additively manufactured Fe-Si. Addit. Manuf. **29**, 100781 (2019)
29. Raissi, M., Perdikaris, P., Karniadakis, G.E.: Physics-informed neural networks: a deep learning framework for solving forward and inverse problems involving nonlinear partial differential equations. J. Comput. Phys. **378**, 686–707 (2019)
30. Roberts, C., Bourell, D., Watt, T., Cohen, J.: A novel processing approach for additive manufacturing of commercial aluminum alloys. Phys. Procedia **83**, 909–917 (2016)
31. Schneidereit, T., Breuß, M.: Collocation polynomial neural forms and domain fragmentation for initial value problems. Neural Comput. Applic. **34**(9), 7141–7156 (2021). arXiv:2103.15413
32. Schneidereit, T., Breuß, M.: Polynomial neural forms using feedforward neural networks for solving differential equations. In: Artificial Intelligence and Soft Computing, Lecture Notes in Computer Science, vol. 12854, pp. 236–245. Springer International Publishing, New York (2021)
33. Schwendner, K., Banerjee, R., Collins, P., Brice, C., Fraser, H.: Direct laser deposition of alloys from elemental powder blends. Scr. Mater. **45**, 1123–1129 (2001)
34. Shoji Aota, L., Bajaj, P., Zschommler Sandim, H., Aimé Jägle, E.: Laser powder-bed fusion as an alloy development tool: parameter selection for in-situ alloying using elemental powders. Materials **13**, 3922 (2020)
35. Speight, J.: Chapter 2—Materials of Construction for Refinery Units. In: Oil And Gas Corrosion Prevention, pp. 3–37 (2014)
36. Strauch, A., Hardes, C., Röttger, A., Uhlenwinkel, V., Baqerzadeh Chehreh, A., Theisen, W., Walther, F., Zoch, H.: Laser additive manufacturing of hot work tool steel by means of a starting powder containing partly spherical pure elements and ferroalloys. Procedia CIRP **94**, 46–51 (2020)
37. Szekely, J., Carlsson, G., Helle, L.: Ladle Metallurgy. Springer Science, Business Media, New York (2012)
38. Taylor, H.M., Karlin, S.: An Introduction to Stochastic Modeling. Academic Press, New York (1998)

39. Tian, P., Ma, J., Zhang, D.: Application of the simulated annealing algorithm to the combinatorial optimisation problem with permutation property: an investigation of generation mechanism. Eur. J. Oper. Res. **118**, 81–94 (1999)
40. Tomellini, M.: Mean field rate equation for diffusion-controlled growth in binary alloys. J. Alloys Compd. **348**(1–2), 189–194 (2003)
41. Tsunoyama, K.: Metallurgy of Ultra-Low-C interstitial-free sheet steel for automobile applications. Phys. Status Solidi (a) **167**, 427–433 (1998)
42. Wang, Y., Karasev, A., Park, J., Jönsson, P.: Non-metallic inclusions in different ferroalloys and their effect on the steel quality: a review. Metall. Mater. Trans. B **52**, 2892–2925 (2021)
43. Wei, G., Huang, P., Xu, C., Liu, D., Ju, X., Du, X., Xing, L., Yang, Y.: Thermophysical property measurements and thermal energy storage capacity analysis of aluminum alloys. Sol. Energy **137**, 66–72 (2016)
44. Wu, L., Liu, K., Zhou, Y.: The Kinetics of Phase Transition of Austenite to Ferrite in Medium-Carbon Microalloy Steel. Metals **11**, 1986–1986 (2021)

MIX
Papier aus verantwortungsvollen Quellen
Paper from responsible sources
FSC® C105338

If you have any concerns about our products,
you can contact us on
ProductSafety@springernature.com

In case Publisher is established outside the EU,
the EU authorized representative is:
Springer Nature Customer Service Center GmbH
Europaplatz 3, 69115 Heidelberg, Germany

Printed by Libri Plureos GmbH
in Hamburg, Germany